W9-AUR-807

TIMEFULNESS

# TIMEFULNESS

*How Thinking Like a Geologist Can Help Save the World*

MARCIA BJORNERUD

PRINCETON UNIVERSITY PRESS
PRINCETON AND OXFORD

Published by Princeton University Press
41 William Street, Princeton, New Jersey 08540

In the United Kingdom: Princeton University Press
6 Oxford Street, Woodstock, Oxfordshire OX20 1TR

press.princeton.edu

Jacket image: Mineralogy lithographs from the *Iconographic Encyclopaedia of Science, Literature, and Art*, 1852



ISBN 978-0-691-18120-2

Library of Congress Control Number: 2018945515

British Library Cataloging-in-Publication Data is available

This book has been composed in Adobe Text Pro and Gotham

Printed on acid-free paper. ∞

Printed in the United States of America

1 3 5 7 9 10 8 6 4 2

## CONTENTS

| | | |
|---|---|---|
| | *Acknowledgments* | vi |
| | Prologue: The Allure of Timelessness | 1 |
| 1 | A Call for Timefulness | 6 |
| 2 | An Atlas of Time | 21 |
| 3 | The Pace of the Earth | 62 |
| 4 | Changes in the Air | 93 |
| 5 | Great Accelerations | 126 |
| 6 | Timefulness, Utopian and Scientific | 159 |
| | Epilogue | 180 |
| | APPENDIXES | |
| I | Simplified Geologic Timescale | 184 |
| II | Durations and Rates of Earth Phenomena | 186 |
| III | Environmental Crises in Earth's History: Causes and Consequences | 190 |
| | *Notes* | 193 |
| | *Index* | 203 |

# ACKNOWLEDGMENTS

I am grateful to the many people who played a part in the evolution of this book: my colleagues David McGlynn and Jerald Podair; Princeton University Press editors Eric Henney and Leslie Grundfest, and associates Arthur Werneck and Stephanie Rojas; copyeditor Barbara Liguori; and illustrator Haley Hagerman, whose work is timeless. Thanks also to my family—my parents, Gloria and Jim; sons, Olav, Finn, and Karl; and beau, Paul, with whom I am lucky to spend my time on Earth.

# TIMEFULNESS

PROLOGUE

# THE ALLURE OF TIMELESSNESS

Time is the one thing we can all agree to call supernatural.

—HALDOR LAXNESS, *UNDER THE GLACIER*, 1968

For children who grow up in wintry climates, few experiences in life will ever elicit the same pure joy as a Snow Day. Unlike holidays, whose pleasures can be diminished by weeks of anticipation, snow days are undiluted serendipities. In the 1970s in rural Wisconsin, school closings were announced on local AM radio, and we'd listen with the volume turned way up, trembling with hope, as the names of public and parochial schools around the county were read, with maddening deliberation, in alphabetical order. At last our school would be named, and in that moment anything seemed possible. Time was temporarily repealed; the oppressive schedules of the adult world magically suspended in a concession to the greater authority of nature.

The day stretched luxuriously before us. An expedition into the white, muted world would be first. We would marvel at the new geography of the woods around the house and the inflation of familiar objects into puffy caricatures of themselves. Stumps and boulders had been fitted with thick cushions; the mailbox wore a ridiculously tall hat. We relished these heroic reconnaissance missions all the more knowing they would be followed by a return to the cozy warmth of the house.

I remember one particular snow day when I was in the eighth grade, that liminal stage when one has access to the realms of both childhood and adulthood. Almost a foot of snow had fallen in the night, followed by fierce winds and biting cold. In the morning, the world was utterly still and blindingly bright. My childhood companions were teenagers now, more interested in sleep than snow, but I could not resist the prospect of a transformed world. I bundled myself in down and wool and stepped outside. The air felt sharp in my lungs. Trees creaked and groaned in that peculiar way that signals deep cold. Trudging down the hill toward the stream below our house, I spotted a dab of red on a branch: a male cardinal huddled in the heatless sunshine. I walked toward the tree and was surprised that the bird didn't seem to hear me. I drew closer still and then realized with repulsion and fascination that it was frozen on its perch in life position, like a glass-eyed specimen in a natural history museum. It was as if time had stopped in the woods, allowing me to see things that were normally a blur of motion.

Back inside that afternoon, savoring the gift of unallocated time, I heaved our big world atlas off the shelf and lay sprawled on the floor with it. I've always been drawn to maps; good ones are labyrinthine texts that reveal hidden histories. On this day, I happened to open the atlas to a two-page chart showing the boundaries of time zones around the globe—the kind with clocks running across the top, showing the relative hour in Chicago, Cairo, Bangkok. The pastel colors on the map ran in mostly longitudinal stripes except for some elaborate gerrymanders like China (all one time zone) and a few outliers, including Newfoundland, Nepal, and central Australia, where the clocks are set ahead or back relative to Greenwich Mean Time by some odd noninteger amount. There were also a few places—Antarctica, Outer Mongolia, and an Arctic archipelago called Svalbard—that were colored gray, which, according to the map legend, signified "No Official Time." I was captivated by the idea of places that had resisted being shackled by measures of time—no minutes or hours, wholly exempt from the tyranny of a schedule. Was time there frozen like the cardinal on the branch? Or simply flowing, unmetered and unfettered, according to a wilder natural rhythm?

Years later, when, through coincidence or predestination, I ended up doing field work for my PhD in geology on Svalbard, I discovered that in some ways, it was indeed a place beyond or outside time. The Ice Age had not yet loosened its grip. Relics of human history from disparate eras—whale bones discarded by seventeenth-century blubber renderers, graves of Russian hunters from the reign of Catherine the Great, the torn fuselage of a Luftwaffe bomber—lay strewn across great barren swaths of tundra as if in a poorly curated exhibition. I also learned that Svalbard's "No Official Time" designation was actually due to a petty, long-running argument between the Russians and Norwegians about whether to observe Moscow or Oslo time there. But on that long-ago snow day, liberated temporarily from quotidian routines, on the cusp of adulthood yet still snug in my parents' house, I had glimpsed the possibility that there were pockets where time remained undefined, amorphous—where one might even travel between past and present with equal freedom. With a dim premonition of the changes and losses that lay ahead, I wished that that perfect day could be my permanent home, from which I might venture but always return to find everything unchanged. This was the start of a complicated relationship with time.

I first traveled to Svalbard as a new graduate student—more specifically, as a seasick passenger aboard a Norwegian Polar Institute research ship—in the summer of 1984. Our field season could not begin until early July, when the sea ice had broken up enough for safe navigation. Three long days after leaving mainland Norway, we at last reached the southwest coast of the island of Spitsbergen, the area that would be the focus of my doctoral work on the tectonic history of the mountain range there, the northernmost extension of the Appalachian-Caledonian chain. In my miserable state of mal de mer, I was actually happy that the waves were too high that day for our small group to be carried to land by rubber boat, because it meant we'd have the luxury of a much quicker, drier trip by helicopter. We flew from the top deck of the pitching ship, with all our gear and food slung like a bag of onions in a net under the helicopter and hanging perilously over the heaving seas. As we approached land I remember searching the ground for some object to provide a sense of scale, but the boulders, streams, and patches of mossy tundra were of indeterminate size. Finally, I saw what looked like a weather-beaten wooden fruit crate. It turned out to be the hut we would live in for the next two months (see figure 1).

Once the helicopter had left and the ship had vanished over the horizon, our camp became detached from late twentieth century. The hut, or *hytte*, which was actually quite snug, had been built from driftwood by resourceful hunters in the early 1900s. We carried World War II–vintage bolt-action Mausers as protection against polar bears. We had no way to communicate with the world apart from a prearranged nightly radio check-in with the ship, which would slowly circumnavigate the archipelago taking oceanographic measurements over the course of the summer. We heard no news about current affairs; for years after that summer and the field seasons that followed, I would discover embarrassing lacunae in my knowledge of world events that had happened between July and September (What? When did Richard Burton die?).

On Svalbard, my perception of time becomes unmoored from the normal measures. It is partly the 24-hour summer daylight (not to say actual sunshine—the weather can be quite

FIGURE 1. The hut on Svalbard, Norwegian arctic

awful), which provides no cue for sleep. But it is also the single-minded focus on the natural history of an austere world that has so little memory of humans. Just as the size of objects is difficult to judge on the tundra, the temporal space between past events becomes hard to discern. The few human-made artifacts one finds—a tangled fishing net, a decaying weather balloon—seem older and shabbier than the ancient mountains, which are robust and vital. Lost in my thoughts on the long walks back to camp each day, my mind washed clean by the sound of wind and waves, I have sometimes felt as if I stood at the center of a circle, equidistant from all stages of my life, past and future. The sensation spills over to the landscape and rocks; immersed in their stories, I see that the events of the past are still present and feel they could even be replayed again one day in a beautiful revelation. This impression is a glimpse not of time*less*ness but time*ful*ness, an acute consciousness of how the world is made by—indeed, made of—time.

CHAPTER 1

# A CALL FOR TIMEFULNESS

*Omnia mutantur, nihil interit* (Everything changes, nothing perishes).

—OVID, *METAMORPHOSES*, AD 8

## A BRIEF HISTORY OF TIME DENIAL

As a geologist and professor I speak and write rather cavalierly about eras and eons. One of the courses I routinely teach is "History of Earth and Life," a survey of the 4.5-billion-year saga of the entire planet—in a 10-week trimester. But as a human, and more specifically as a daughter, mother, and widow, I struggle like everyone else to look Time honestly in the face. That is, I admit to some time hypocrisy.

Antipathy toward time clouds personal and collective thinking. The now risible "Y2K" crisis that threatened to cripple global computer systems and the world economy at the turn of the millennium was caused by programmers in the 1960s and '70s who apparently didn't really think the year 2000 would ever arrive. Over the past decade, Botox treatments and plastic surgery have come to be viewed as healthy boosts to self-esteem rather than what they really are: evidence that we fear and loathe our time-iness. Our natural aversion to death is amplified in a culture that casts Time as an enemy and does everything it can to deny its passage. As Woody Allen said: "Americans believe death is optional."

This type of time denial, rooted in a very human combination of vanity and existential dread, is perhaps the most common and forgivable form of what might be called *chronophobia*. But there are other, more toxic varieties that work together with the mostly benign kind to create a pervasive, stubborn, and dangerous temporal illiteracy in our society. We in the twenty-first century would be shocked if an educated adult were unable to identify the continents on a world map, yet we are quite comfortable with widespread obliviousness about anything but the most superficial highlights from the planet's long history (um, Bering Strait . . . dinosaurs . . . Pangaea?). Most humans, including those in affluent and technically advanced countries, have no sense of temporal proportion—the *durations* of the great chapters in Earth's history, the *rates* of change during previous intervals of environmental instability, the *intrinsic timescales* of "natural capital" like groundwater systems. As a species, we have a childlike disinterest and partial disbelief in the time before our appearance on Earth. With no appetite for stories lacking human protagonists, many people simply can't be bothered with natural history. We are thus both intemperate and intemp*o*rate—time illiterate. Like inexperienced but overconfident drivers, we accelerate into landscapes and ecosystems with no sense of their long-established traffic patterns, and then react with surprise and indignation when we face the penalties for ignoring natural laws. This ignorance of planetary history undermines any claims we may make to modernity. We are navigating recklessly toward our future using conceptions of time as primitive as a world map from the fourteenth century, when dragons lurked around the edges of a flat earth. The dragons of time denial still persist in a surprising range of habitats.

Among the various foes of time, Young Earth creationism breathes the most fire but is at least predictable in its opposition.

In years of teaching geology at the university level, I have had students from evangelical Christian backgrounds who earnestly struggle to reconcile their faith with the scientific understanding of the Earth. I truly empathize with their distress and try to point out paths toward resolution of this internal discord. First, I emphasize that my job is not to challenge their personal beliefs but to teach the logic of geology (geo-logic?)—the methods and tools of the discipline that enable us not only to comprehend how the Earth works at present but also to document in detail its elaborate and awe-inspiring history. Some students seem satisfied with keeping science and religious beliefs separate through this methodological remove. But more often, as they learn to read rocks and landscapes on their own, the two worldviews seem increasingly incompatible. In this case, I use a variation on the argument made by Descartes in his *Meditations* about whether his experience of Being was real or an elaborate illusion created by a malevolent demon or god.[1]

Early in an introductory geology course, one begins to understand that rocks are not nouns but verbs—visible evidence of processes: a volcanic eruption, the accretion of a coral reef, the growth of a mountain belt. Everywhere one looks, rocks bear witness to events that unfolded over long stretches of time. Little by little, over more than two centuries, the local stories told by rocks in all parts of the world have been stitched together into a great global tapestry—the geologic timescale. This "map" of Deep Time represents one of the great intellectual achievements of humanity, arduously constructed by stratigraphers, paleontologists, geochemists, and geochronologists from many cultures and faiths. It is still a work in progress to which details are constantly being added and finer and finer calibrations being made. So far, no one in more than 200 years has found an anachronistic rock or fossil—as biologist J.B.S. Haldane reputedly said, "a Precambrian rabbit"[2]—that would represent a fatal internal inconsistency in the logic of the timescale.

If one acknowledges the credibility of the methodical work by countless geologists from around the world (many in the service of petroleum companies), and one believes in a God as creator, the choice is then whether to accept the idea of (1) an ancient and complex Earth with epic tales to tell, set in motion eons ago by a benevolent creator, or (2) a young Earth fabricated only a few thousand years ago by a devious and deceitful creator who planted specious evidence of an old planet in every nook and cranny, from fossil beds to zircon crystals, in anticipation of our explorations and laboratory analyses. Which is more heretical? A corollary of this argument, to be deployed with tact and care, is that compared with the deep, rich, grand geologic story of Earth, the Genesis version is an offensive dumbing-down, an oversimplification so extreme as to be disrespectful to the Creation.

While I have sympathy for individuals wrestling with theological questions, I have no tolerance for those who intentionally spread brain-fogging pseudoscience under the aegis of (suspiciously well-funded) religious organizations. My colleagues and I despair at the existence of atrocities like Kentucky's Creation Museum, and the disheartening frequency with which Young Earth websites appear when students search for information about, say, isotopic dating. But I hadn't fully understood the tactics and far-reaching tentacles of the "Creation Science" industry until a former student alerted me that one of my own papers, published in a journal read only by nerdy geophysicists, had been cited on the website of the Institute for Creation Research. Citation frequency is one metric by which the scientific world ranks its practitioners, and most scientists adopt P. T. Barnum's view that there is "no such thing as bad publicity"—the more citations, the better, even if one's ideas are being rebutted or challenged. But this citation was akin to a social media endorsement from an especially despised troll.

The article was about some unusual metamorphic rocks in the Norwegian Caledonides whose high-density minerals attest to their having been at crustal depths of at least 50 km (30 mi) at the time the mountain belt was forming. Oddly, these rocks occur in lenses and pods, interleaved with rock masses that did not undergo the conversion to the more compact mineral forms. My coinvestigators and I showed that the nonuniform metamorphism was due to the extremely dry nature of the original rocks, which inhibited the recrystallization process. We argued that the rocks, with their low-density minerals, probably resided unstably for some period in the deep crust until one or more large earthquakes fractured the rocks and allowed fluids to enter and locally trigger long-suppressed metamorphic reactions. We used some theoretical constraints to suggest that in this case, the spotty metamorphism might have happened in thousands or tens of thousands of years, rather than the hundreds of thousands to millions of years in more typical tectonic settings. This "evidence for rapid metamorphism" is what someone at the Institute for Creation Research grabbed onto and cited—completely ignoring the fact that the rocks are known to be about a billion years old and that the Caledonides were formed around 400 million years ago. I was stunned to realize that there are people with enough time, training, and motivation to be trawling the vast waters of the scientific literature for such finds, and that someone is probably paying them to do it. The stakes must be very high.

For those who deliberately confuse the public with falsified accounts of natural history, colluding with powerful religious syndicates to promote doctrine that serves their own coffers or political agendas, my Midwestern niceness reaches its limit. I would love to say: "No fossil fuels for you (or plastic, for that matter). All that oil was found thanks to a rigorous understanding of the sedimentary record of geologic time. And no modern medicine for you either, since the great majority of pharmaceutical, therapeutic, and surgical advances involve testing on mice, which makes sense only if you understand that they are our evolutionary kin. You can cleave to whatever myths you like about the history of the planet, but then you should live with only the technologies that follow from that worldview. And please stop dulling the minds of the next generation with retrograde thinking." (Wow! I feel better now.)

Some religious sects embrace a symmetrical form of time denial, believing not only in a truncated geologic past but also a foreshortened future in which the Apocalypse is nigh. Fixation with the end of the world may seem a harmless delusion—the lone robed man with a warning placard is a cartoon cliché, and we've all come through several "Rapture" dates unscathed. But if enough voters truly think this way, there are serious policy implications. Those who believe that the End of Days is just around the corner have no reason to be concerned about matters like climate change, groundwater depletion, or loss of biodiversity.[3] If there is no future, conservation of any kind is, paradoxically, wasteful.

As exasperating as professional Young-Earthers, creationists, and apocalypticists can be, they are completely forthright about their chronophobia. More pervasive and corrosive are the nearly invisible forms of time denial that are built into the very infrastructure of our society. For example, in the logic of economics, in which labor productivity must always increase to justify higher wages, professions centered on tasks that simply take time—education, nursing, or art performance—constitute a problem because they cannot be made significantly more efficient. Playing a Haydn string quartet takes just as long in the twenty-first century as it did in the eighteenth; no progress has been made! This is sometimes called "Baumol's disease" for one of the economists that first described the dilemma.[4] That it is considered a pathology reveals much about our attitude toward time and the low value we in the West place on process, development, and maturation.

Fiscal years and congressional terms enforce a blinkered view of the future. Short-term thinkers are rewarded with bonuses and reelection, while those who dare to take seriously our responsibility to future generations commonly find themselves outnumbered, outshouted, and out of office. Few modern public entities are able to make plans beyond biennial budget cycles. Even two years of forethought seems beyond the capacity of Congress and state legislatures these days, when last-minute, stop-gap spending measures have become the norm. Institutions that do aspire to the long view—state and national parks, public libraries, and universities—are increasingly seen as taxpayer burdens (or untapped opportunities for corporate sponsorship).

Conserving natural resources—soil, forests, water—for the nation's future was once considered a patriotic cause, evidence of love of country. But today, consumption and monetization have become strangely mixed up with the idea of good citizenship (a concept that now includes corporations). In fact, the word *consumer* has become more or less a synonym for *citizen*, and that doesn't really seem to bother anyone. "Citizen" implies engagement, contribution, give-and-take. "Consumer" suggests only taking, as if our sole role is to devour everything in sight, in the manner of locusts descending on a field of grain. We might scoff at apocalyptic thinking, but the even more pervasive idea—indeed, economic credo—that levels of consumption can and should increase continuously is just as deluded. And while the need for long-range vision grows more acute, our attention spans are shrinking, as we text and tweet in a hermetic, narcissistic Now.

Academe, too, must take some responsibility for promulgating a subtle strain of time denial in the way that it privileges certain types of inquiry. Physics and chemistry occupy the top echelons in the hierarchy of intellectual pursuits owing to their quantitative exactitude. But such precision in characterizing how nature works is possible only under highly controlled, wholly unnatural conditions, divorced from any particular history or moment. Their designation as the "pure" sciences is revealing; they are pure in being essentially atemporal—unsullied by time, concerned only with universal truths and eternal laws.[5] Like Plato's "forms," these immortal laws are often considered more real than any specific manifestation of them (e.g., the Earth). In contrast, the fields of biology and geology occupy lower rungs of the scholarly ladder because they are very "impure," lacking the heady overtones of certainty because they are steeped through and through with time. The laws of physics and chemistry obviously apply to life-forms and rocks, and it is also possible to abstract some general principles about how biological and geologic systems function, but the heart of these fields lies in the idiosyncratic profusion of organisms, minerals, and landscapes that have emerged over the long history of this particular corner of the cosmos.

Biology as a discipline is elevated by its molecular wing, with its white-coat laboratory focus and its venerable contributions to medicine. But lowly geology has never achieved the glossy prestige of the other sciences. It has no Nobel Prize, no high school Advanced Placement courses, and a public persona that is musty and dull. This of course rankles geologists, but it also has serious consequences for society at a time when politicians, CEOs, and ordinary citizens urgently need to have some grasp of the planet's history, anatomy, and physiology.

For one thing, the perceived value of a science profoundly influences the funding it receives. Out of frustration with limited grant money for basic geologic investigations, some geochemists and paleontologists studying the early Earth and the most ancient traces of life in the rock record have cleverly recast themselves as "astrobiologists" to ride on the coattails of NASA initiatives that support research into the possibility of life elsewhere in the Solar System or beyond. While I admire this shrewd maneuver, it is disheartening that we geologists must wrap ourselves in the hype of the space program to make legislators or the public interested in their own planet.

Second, the ignorance of and disregard for geology by scientists in other fields has serious environmental consequences. The great advances in physics, chemistry, and engineering made in the Cold War years—development of nuclear technologies; synthesis of new plastics, pesticides, fertilizers, and refrigerants; mechanization of agriculture; expansion of highways—ushered in an era of unprecedented prosperity but also left a dark legacy of groundwater contamination, ozone destruction, soil and biodiversity loss, and climate change for subsequent generations to pay for. To some extent, the scientists and engineers behind these achievements can't be blamed; if one is trained to think of natural systems in highly simplified ways, stripping away the particulars so that idealized laws apply, and one has no experience with how perturbations to these systems may play out over time, then the undesirable consequences of these interventions will come as a surprise. And to be fair, until the 1970s, the geosciences themselves did not have the analytical tools with which to conceptualize the behavior of complex natural systems on decade to century timescales.

By now, however, we should have learned that treating the planet as if it were a simple, predictable, passive object in a controlled laboratory experiment is scientifically inexcusable. Yet the same old time-blind hubris is allowing the seductive idea of climate engineering, sometimes called geoengineering, to gain traction in certain academic and political circles. The most commonly discussed method for cooling the planet without having to do the hard work of cutting greenhouse gas emissions is the injection of reflective sulfate aerosol particles into the stratosphere—the upper atmosphere—to mimic the effect of large volcanic eruptions, which have cooled the planet temporarily in the past. The 1991 eruption of Mount Pinatubo in the Philippines, for example, caused a two-year pause in the steady climb of global temperatures. The chief advocates for this type of planetary tinkering are physicists and economists, who argue that it would be cheap, effective, and technologically feasible, and promote it under the benign, almost bureaucratic-sounding name "Solar Radiation Management."[6]

But most geoscientists, acutely aware of how even small changes to intricate natural systems can have large and unanticipated consequences, are profoundly skeptical. The volumes of sulfate required to reverse global warming would be equivalent to a Pinatubo-sized eruption every few years—for at least the next century—since halting the injections in the absence of significant reduction in greenhouse gas levels would result in an abrupt global temperature spike that might be beyond the adaptive capacity of much of the biosphere. Even worse, the effectiveness of the approach wanes with time, because as stratospheric sulfate concentrations increase, the tiny particles coalesce into larger ones, which are less reflective and have a shorter residence time in the atmosphere. Most important, even though there would probably be a net decrease in overall global temperature, we have no way of knowing exactly how regional or local weather systems would be affected. (And by the way, we have no international governance mechanism to oversee and regulate planetary-scale manipulation of the atmosphere).

In other words, it is time for all the sciences to adopt a geologic respect for time and its capacity to transfigure, destroy, renew, amplify, erode, propagate, entwine, innovate, and exterminate. Fathoming deep time is arguably geology's single greatest contribution to humanity. Just as the microscope and telescope extended our vision into spatial realms once too minuscule or too immense for us to see, geology provides a lens through which we can witness time in a way that transcends the limits of our human experiences.

But even geology cannot exempt itself from culpability for public misconceptions about time. Since the birth of the discipline in the early 1800s, geologists—congenitally wary of Young-Earthers—have droned on about the unimaginable slowness of geologic processes, and the idea that geologic changes accrue only over immense periods of time. Moreover, geologic textbooks invariably point out (almost gleefully) that if the 4.5 billion-year story of the Earth is scaled to a 24-hour day, all of human history would transpire in the last fraction of a second before midnight. But this is a wrongheaded, and even irresponsible, way to understand our place in Time. For one thing, it suggests a degree of insignificance and disempowerment that not only is psychologically alienating but also allows us to ignore the magnitude of our effects on the planet in that quarter second. And it denies our deep roots and permanent entanglement with Earth's history; our specific clan may not have shown up until just before the clock struck 12:00, but our extended family of living organisms has been around since at least 6 a.m. Finally, the analogy implies, apocalyptically, that there is no future—what happens after midnight?

## A MATTER OF TIME

While we humans may never completely stop worrying about time and learn to love it (to borrow a turn of phrase from *Dr. Strangelove*), perhaps we can find some middle ground between chronophobia and chronophilia, and develop the habit of timefulness—a clear-eyed view of our place in Time, both the past that came long before us and the future that will elapse without us.

Timefulness includes a feeling for distances and proximities in the geography of deep time. Focusing simply on the age of the Earth is like describing a symphony in terms of its total measure count. Without time, a symphony is a heap of sounds; the durations of notes and recurrence of themes give it shape. Similarly, the grandeur of Earth's story lies in the gradually unfolding, interwoven rhythms of its many movements, with short motifs scampering over tones that resonate across the entire span of the planet's history. We are learning that the tempo of many geologic processes is not quite as *larghissimo* as once thought; mountains grow at rates that can now be measured in real time, and the quickening pace of the climate system is surprising even those who have studied it for decades.

Still, I am comforted by the knowledge that we live on a very old, durable planet, not an immature, untested, and possibly fragile one. And my daily experience as an earthling is enriched by an awareness of the lingering presence of so many previous versions and denizens of this place. Understanding the reasons for the morphology of a particular landscape is similar to the rush of insight one has upon learning the etymology of an ordinary word. A window is opened, illuminating a distant yet recognizable past—almost like remembering something long forgotten. This enchants the world with layers of meaning and changes the way we perceive our place in it. Although we may fervently wish to deny time for reasons of vanity, existential angst, or intellectual snobbery, we diminish ourselves by denouncing our temporality. Bewitching as the fantasy of timelessness may be, there is far deeper and more mysterious beauty in timefulness.

## A SHORT LOOK AHEAD

I've written this book in the belief (possibly naïve) that if more people understood our shared history and destiny as Earth-dwellers, we might treat each other, and the planet, better. At a time when the world appears more deeply divided than ever by religious dogmas and political animosities, there would seem to be little hope of finding a common philosophy or list of principles that might bring all factions to the table for honest discourse about increasingly intractable environmental, social, and economic problems.

But the communal heritage of geology may yet allow us to reframe our thinking about these issues in a fresh new way. In fact, natural scientists already serve as a kind of impromptu international diplomatic corps who demonstrate that it is possible for people from developed and developing countries, socialist and capitalist regimes, theocracies and democracies to co-operate, debate, disagree, and move toward consensus, unified by the fact that we are all citizens of a planet whose tectonic, hydrologic, and atmospheric habits ignore national boundaries. Maybe, just maybe, the Earth itself, with its immensely deep history can provide a politically neutral narrative from which all nations may agree to take counsel.

In the chapters that follow, I hope to convey the mind-altering sense of time and planetary evolution that permeates geologic thinking. It may not be possible to grasp fully the immensity of geologic time, but one can at least develop some feeling for its proportions. I once had a math professor who was fond of reminding the class that "there are many sizes and shapes of infinity." Something similar can be said about geologic time, which though not actually infinite is effectively so from a human perspective. But there are different depths in the seas of Deep Time—from the shallows of the last Ice Age to the abyss of the Archean. Chapter 2 tells the story of how geologists mapped the ocean of time, first qualitatively using the fossil record, then with increasing quantitative precision through the phenomenon of natural radioactivity. (This is the most technical material in the book; if isotope geochemistry just isn't your thing, you can skip the details and move on without guilt or loss of continuity). The geologic timescale is an underappreciated collaborative intellectual achievement, and still a work in progress. A simplified version of the timescale is provided for reference in appendix I.

Chapter 3 is about the intrinsic rhythms of the solid earth—the paces of tectonics and landscape evolution, and how a geologic perspective requires us to abandon any belief in the permanence of topographic features. Geologic processes may be slow, but they are not beyond our perception. And one of the most important insights to emerge from "clocking the Earth" is that the rates of disparate natural processes, from the growth of mountains, to erosion, to evolutionary adaptation—each powered by different motive forces—are remarkably well matched. The durations, rates, and recurrence intervals of various geologic phenomena are summarized in several tables in appendix II.

Chapter 4 is about the evolution of the atmosphere and the rates of change in its composition during environmental upheavals and mass extinctions in geologic history. A recurrent theme is that long periods of planetary stability have ended abruptly in the past when rates of environmental change outpaced the biosphere's capacity to adapt (and in only one case can we lay the blame on a meteorite). Appendix III compares the causes and consequences of eight great environmental crises in Earth's history, including changes unfolding now.

Chapter 5 begins with the discovery of the Ice Age (the Pleistocene) in the nineteenth century and explains how modern understanding of climate change gradually emerged from that. The Pleistocene was not simply an interval of constant cold, but more than 2 million years of climate variability. It was the transition into the climatically stable Holocene 10,000 years ago that allowed the emergence of modern human civilization. This is sobering in light of current rates of environmental change, which are virtually unprecedented in geologic time—the basis for the argument that we are now in a new geologic epoch, the Anthropocene.

The final chapter looks to the geologic future and outlines ideas for building a more robust, enlightened, time-literate society that is able to make decisions on intergenerational timescales. This requires only a shift in perception. For many in North America, the 2017 total solar eclipse was a transformative experience, a fleeting vision of our place in the cosmos. Similarly, geologic observation provides a view of the strange and scintillating world of Time we dwell in but cannot ordinarily see. Even a glimpse can alter one's experience of being alive on Earth.

## CHAPTER 2

# AN ATLAS OF TIME

> Although we are mere sojourners on the surface of the planet, chained to a mere point in space, enduring but for a moment of time, the human mind is not only enabled to number worlds beyond the unassisted ken of mortal eye, but to trace the events of indefinite ages before the creation of our race.
>
> —CHARLES LYELL, *PRINCIPLES OF GEOLOGY*, 1830

### THINKING LIKE A ROCK

Like many geologists, I stumbled into the discipline more or less by accident. Geoscience is not present or prioritized in most U.S. high school curricula in the same way that physics, chemistry, and biology are, and as a result, there are few students who enter university aware of geology as a mature academic field with its own lively intellectual culture. As a first-year college student with a proclivity for the humanities, I enrolled in an introductory geology course mainly to fulfill a science requirement. My expectations were rather low; it was "rocks for jocks." The weekly field trips would at least be a chance to get off campus. To my surprise, I found that geology demanded a type of whole-brain thinking I hadn't encountered before. It creatively appropriated ideas from physics and chemistry for the investigation of unruly volcanoes and oceans and ice sheets. It applied scholarly habits one associates with the study of literature and the arts—the practice of close reading, sensitivity to allusion and analogy, capacity for spatial visualization—to the examination of rocks. Its particular form of inferential logic demanded mental versatility and a vigorous but disciplined imagination. And its explanatory power was vast; it was nothing less than the etymology of the world. I was hooked.

An apt way to describe how geologists perceive rocks and landscapes is the metaphor of a *palimpsest*—the term used by medieval scholars to describe a parchment that was used more than once, with old ink scraped off to allow a new document to be inscribed. Invariably, the erasure was imperfect, and vestiges of the earlier text survived. These remnants can be read using x-rays and various illumination techniques, and in some cases are the only sources of very ancient documents (including several of the most important writings of Archimedes). In the same way, everywhere on Earth, traces of earlier epochs persist in the contours of landforms and the rocks beneath, even as new chapters are being written. The discipline of geology is akin to an optical device for seeing the Earth text in all its dimensions. To think geologically is to hold in the mind's eye not only what is visible at the surface but also present in the subsurface, what has been and will be.

Other disciplines, especially cosmology, astrophysics, and evolutionary biology, are concerned with Deep Time (John McPhee's evocative phrase for the prehistorical, prearcheological past[1]), but geology is unique in having direct access to tangible objects that witnessed it. Geology is not concerned with the nature of time per se but rather with its unmatched powers of transformation. In documenting the evidence for earlier versions of the world, geologists were the first to develop an instinct for the immensity of planetary time, even though they had no way of measuring it until the twentieth century.

## HOW THE EARTH GOT OLD (THEN A LOT YOUNGER)

Among the sciences, geology is something of a late bloomer. The motions of the planets were explained in the seventeenth century, the laws of thermodynamics and electromagnetism were worked out in the nineteenth, and the secrets of the atom were known in the early twentieth, all before we knew the age of the Earth or had any clear idea about its planetary-scale behavior. This does not mean geologists have been dullards but rather that Earth has been an elusive subject to study—simultaneously too near and too far away to get into clear view. When other sciences were making great strides toward describing nature using telescopes, microscopes, beakers, and bell jars, Earth could neither be viewed through one lens nor reduced to a laboratory-sized experiment. Also, interpreting the Earth has always been deeply entangled with our self-perception as humans and our cherished stories about our relationship to the rest of creation. No wonder it is difficult to step back and see things in clear perspective.

More than any other scientific discipline, geology requires prodigious powers of visualization and openness to bold inductive inferences. How, for example, could someone in the eighteenth century begin to answer the question, How old is the Earth? In the Western world, most people had no reason to challenge the 6,000 years or so implied in the Bible (in 1654, the archbishop of the Church of Ireland, James Ussher, with astonishing precision, had calculated the date of the creation: Sunday, 23 October 4004 BC). When I ask twenty-first century students how they would go about answering this question on their own, setting aside religious preconceptions and the 4.5 billion-year figure they have been told, they usually say something like, "Well, find the oldest rocks and figure out how old they are," and then realize this is no answer—how does one know which rocks are oldest, and how does one go about determining their age? One needs the whole edifice of modern geology even to begin. So it is truly extraordinary that in 1789, a

Scottish physician, gentleman farmer, and natural philosopher had the insight to see the vastness of geologic time in an outcrop on the coast near Dunbar.[2]

At a blustery cape called Siccar Point, Hutton noted a discontinuity between two sequences of sedimentary rocks, a surface dividing a lower sequence in which the layers were nearly vertical, from an upper one with the layers closer to the expected horizontal (see figure 2). Many people had seen this promontory before; anyone in a boat would have been careful to steer clear of it to avoid being caught in the waves that crash on the rocks. Hutton, however, was able to see the rocks not merely as a navigational hazard but as a vivid record of vanished landscapes. He made two astoundingly perceptive interpretive leaps. First, he recognized that the underlying vertical rocks represented a former mountain range where marine strata had been tilted by crustal upheaval. Second, he understood that the surface that truncated them represented an erosional interval long enough to wear down the mountains, and that the overlying rocks were sediments that had accumulated on top of their ruins.

Based on his estimate of the rate of erosion on his own land, Hutton asserted that the discontinuity—now called an angular unconformity—represented an unfathomably long interval of time, essentially infinite compared with the biblically ordained age of the Earth. In this simple but revolutionary calculation, Hutton broke with the prevailing belief that Earth's present and past were governed by different regimes, that a violent past of cataclysms like Noah's flood had given way to the unchanging world of the present. Under the assumption that the Earth was only a few thousand years old, deeply eroded valleys and thick piles of sedimentary rock could be explained only by large-magnitude catastrophic events. Hutton replaced this worldview with the foundational idea of geology: *uniformitarianism*—the

FIGURE 2. Hutton's unconformity at Siccar Point, Scotland

assumption that present-day processes are the same as those that operated in the geologic past.

But Hutton's geologic imagination went still further. In his 1789 treatise *Theory of the Earth*, he made the even more daring generalization that this particular unconformity recorded just one iteration in an endless cycle of rock accumulation, uplift, erosion, and renewal on Earth, extending backward into the dim mists of time. Hutton's singular intuition about Deep Time—a radical rewriting of Earth's past—opened the intellectual doors through which modern geology and biology could emerge. Without Hutton and his champion, Charles Lyell, who raised uniformitarianism to orthodoxy a generation later in his massive, rhetorically virtuosic *Principles of Geology*, Charles Darwin would not have had his insight about the power of time to shape organisms through natural selection. (Lyell's exhortations about the antiquity of the Earth echoed in Darwin's head during his five years on the HMS *Beagle* expedition; the first volume of *Principles of Geology* was perhaps the most important book in the small library he brought with him). But Hutton's appealing vision of a world in an infinite, repeating loop was in some ways a chimera, an abstraction that excused itself from the harder, messier work of reconstructing the particulars of Earth's biography. In Greek, there is a useful distinction between time as something that simply marches on—*chronos*—and time that is defined within a narrative—*kairos*. Hutton gave us the first glimmers of planetary *chronos*, but the task of calibrating it, and adding *kairos*, has consumed geologists for the past two centuries.

Early attempts to transcribe the geologic record into an account of Earth's history were based on the idea that certain rock types had formed worldwide at distinct times in the past. Crystalline rocks like granites and gneisses were considered the original or "Primary" rocks, while stratified ones such as limestones and sandstones were "Secondary." Semicohesive gravel and sand deposits were "Tertiary," and loose, uncemented sediments were "Quaternary" (the latter term persists, quaintly, on the modern geologic timescale, and "Tertiary" survived into the late twentieth century). But there was no basis for knowing whether the ages of these rock varieties were truly the same from place to place.

In the early 1800s, the first preliminary sketches for a well-calibrated chart of deep time were made possible by the astute observations of a canal-digger named William Smith, who noted that certain types of fossil shells occurred in the same sequence in strata all across England (see figure 3). These *index fossils*—as distinctive to specific geologic periods as pillbox hats and bell bottoms are to cultural eras—made it possible

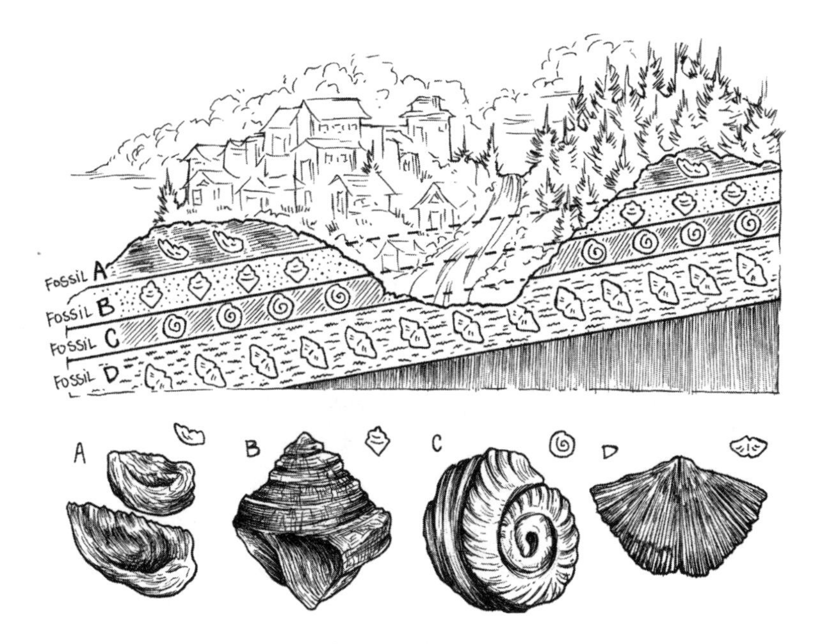


FIGURE 3. The concept of index fossils

to draw connections between layers that are not spatially continuous, first in Britain, then across the Channel into France. Amateur fossil collectors like the celebrated Mary Anning of Lyme Regis—immortalized in the "She sells seashells" tongue twister—were essential to the early stages of assembling the geologic timescale. Old ideas that rock strata were global in nature and recorded the same events worldwide had to be abandoned; the planet's long history turned out to be far more complex than Hutton ever imagined. But decades of laborious mapping and collecting, classifying and cataloging, lumping and splitting ultimately led to the global correlation of sedimentary sequences from all over the world.

The result is the geologic timescale most familiar to the public: going backward from the present, the Cenozoic Era with its multifarious mammals, the Mesozoic and its redoubtable reptiles, the Paleozoic with murky coal-swamps, gasping lungfish, and scuttling trilobites. The rich profusion of fossil life-forms allowed each era to be further subdivided into periods, periods into epochs, epochs into ages. But beneath the lowest shelly layers in the Paleozoic rocks, below strata of the Cambrian Period, the rocks fell silent; no fossils could be found. It appeared that life had sprung suddenly into existence in the Cambrian, a vexing mystery that greatly troubled Darwin. Without visible fossils, the one tool that Victorian geologists had for demarcating geologic time, these oldest rocks were a knotty skein that could not be untangled, so they were simply shelved under the name "Precambrian." It would take a century before geologists would recognize that the Precambrian Earth teemed with life—and that Precambrian time represents almost 90% of Earth's history.

I think of the second half of the nineteenth century as the dark ages for geology.

After Hutton's transcendental vision of a self-renewing Earth, Lyell's inspirational treatise on how the new science of geology would make it possible to "trace the events of indefinite ages," and Darwin's brilliant synthesis of biological and geologic observations, internal and external forces conspired to slow the intellectual momentum. Among these forces was the indomitable physicist William Thomson, Lord Kelvin (1824–1907), who began to take an interest in geology soon after the publication of Darwin's *On the Origin of Species.* As the high priest of thermodynamics, Kelvin rightly attacked the Huttonian idea of an infinitely old Earth—a kind of perpetual motion machine—as a violation of his second law. But his particularly ferocious attack on Darwin for an unsophisticated estimate of the minimum age of the Earth in the first edition of *Origin* suggests that the motivations were not entirely scientific.

Darwin somehow sensed, without any knowledge of the actual mechanism of heredity, that evolution by natural selection would have required hundreds of millions to billions of years to produce the observed diversity of living and fossil life-forms. His intuition about the magnitude of geologic time was truly remarkable, but it was undermined by his inclusion in *Origin* of a single poorly judged attempt at quantification. Like Hutton, Darwin used erosion as a metric of elapsed time. Greatly underestimating the power of English rivers to sculpt the landscape, he suggested it had taken a single valley, the Weald, about 300 million years to form (a value too large by a factor of at least 100). Since the rocks that formed the valley walls were still older, yet among the youngest in the region, Darwin surmised that the Earth itself could be a thousand million (billion) years old or more. His conclusion was—astonishingly—correct, but this one argument, in a book that is otherwise a paragon of carefully wrought exposition, was naïve and easily demolished.

Starting in the early 1860s, Kelvin published a series of papers in which he used the most advanced physics of the day to estimate the age of the Earth based on assumptions about the rate of conductive cooling of the planet and the lifespan of the Sun. Between 1864 and 1897, his determination of Earth's age shrank from a few hundreds of millions of years to just 20 million years. As the time Kelvin would allot to geology continued to contract, a few frustrated geologists attempted to reclaim the question and made independent estimates by summing the thicknesses of all known strata from Cambrian to recent time, then dividing the total by an assumed sedimentation rate. This approach yielded ages of hundreds of millions to billions of years, but the large uncertainties involved made the results easy to dismiss. A small number of younger physicists who were able to follow Kelvin's calculations began to question his framing suppositions—which would indeed be proven wrong decades later—but they were reluctant to incur the wrath of the leading scientist of the day. A brave chemist, John Joly (who would later invent color photography), suggested that the sodium content of seawater was a *proxy*, or stand-in, for the age of the Earth. His (also erroneous) assumption was that the sea had become progressively more saline over time as rivers delivered dissolved elements from rocks on land to the sea. By using typical values of sodium dissolved in river water, Joly estimated Earth's age at 100 million years, gaining back some of the ground geologists had lost to Lord Kelvin.[3]

In his later years, Darwin called Kelvin his "sorest trouble." Darwin died in 1882, haunted by uncertainties about his own lifework, which he felt in his marrow must be correct. Twentieth-century physics would finally rebut Kelvin's arguments, but Kelvin's true purposes were made plain in a speech he made on the occasion of his election as president of the British Association for the Advancement of Science: "I have always felt that the hypothesis of natural selection does not contain the true theory of evolution, if evolution there has been in biology. . . . Overpoweringly strong proofs of intelligent and benevolent design lie around us . . . , and teaching us that all living things depend on one everlasting Creator and Ruler."[4]

## PAUSING FOR TEA WITH CHARLES

The question of the duration of geologic time probably mattered more deeply to Darwin than to any other person in history, and every time I think about the intellectual dissonance he must have suffered in the last decades of his life, I feel a surge of empathy for him. On the 200th anniversary of Darwin's birth, I organized an all-day reading of *On the Origin of Species* at our university library, with dozens of faculty, staff, and students each taking a turn reading aloud for 20-minute stints, with breaks every hour for brief discussion.

The event took place in the period-appropriate, wood-paneled venue of the rare-books room. We served tea and scones with marmalade, and a few people even showed up in Victorian-era dress. Although I knew this would be an intellectually engaging event, I hadn't anticipated it would also be an emotionally moving experience. Over the course of the day, the cumulative effect of hearing Darwin's words spoken aloud was overwhelming. Through the voices of men and women, scientists and musicians, philosophers and economists, young, middle-aged, and old, Darwin's own very human voice could be heard—his delight in the minutest details of the natural world, his earnest thoroughness as a scientist (several people fell asleep during the long sections on pigeon breeding), his personal timidity and reluctant role as a revolutionary, and, most affectingly, his wracking self-doubt and preemptive defensiveness against the attacks that he fully anticipated. *Origin* is a humbly argued, methodical, (and quite often tedious) explication of an idea that Darwin was convinced must be right but also knew would be subject to savage critiques. He did not, however, seem to think the question of geologic time would be one of the scientific objections. In Chapter 9, he wrote: "He who can read Sir Charles Lyell's grand work on the Principles of Geology, which the future historian will recognise as having produced a revolution in natural science, yet does not admit how incomprehensively vast have been the past periods of time, may at once close this volume."

By the end of the reading marathon, it seemed as if Darwin had been in the room with us, and I had a strong, irrational wish to speak with him. I recalled the painting of an elderly Darwin that hangs in the National Portrait Gallery in London.

It depicts a hunched, sad-eyed man who, it seemed to me, was almost physically cramped by the intellectual limits of his day. I yearned to convey to him how marvelously his simple idea has flowered and itself evolved, informing countless new fields of inquiry, and to share with him scientific news that would have eased his troubled mind: Earth *is* old.

## ROCKS KEEP TIME

In addition to the injury done to Darwin, the controversy over the age of the Earth caused lasting damage to geology. When the conclusions of physics seemed incompatible with the increasingly detailed documentation of Earth's long history, some geologists declared that geology had to break with other sciences and pursue its own methods as a wholly independent field of inquiry. This aggravation at the impasse with physics is understandable, but it would unfortunately influence the way generations of geologists were educated, and it set the discipline back by decades. Distaste for physics and distrust of those not trained as geologists contributed, for example, to geology's long, obstinate denial of the evidence for moving continents. In 1915, a German meteorologist, Alfred Wegener, presented carefully documented evidence that Earth's landmasses had once been united in a supercontinent, Pangaea. But Wegener's lack of geologic credentials (combined with American and British antipathy toward all things German during World War I) made his ideas anathema within geologic circles until the plate tectonic revolution of the 1960s.

But in the first years of the twentieth century, a revolution in physics would finally provide the tools to lead geology out of the Victorian labyrinth in which it had become lost. Only a decade after the accidental discovery of the phenomenon of radioactivity by Henri Becquerel in 1897, it would already be used to determine the age of rocks. By 1902, the work of Marie Curie in Paris and Ernest Rutherford at Cambridge had shown that radioactive decay was a kind of natural alchemy in which some elements (for example, uranium) spontaneously emit energy as they transmute to other elements (e.g., lead), at a consistent rate proportional to the remaining amount of the first element. We would now say that certain elements—which are always defined by the number of protons in their nucleus—have different subtypes called *isotopes*, with varying numbers of neutrons, and that some of these *parent* isotopes decay to *daughter* isotopes of other elements. But the structure of the atom was not even known at that time; the nucleus wasn't discovered by Rutherford until 1911, and the concept of isotopes emerged several years after that.

In 1905, Rutherford demonstrated that radioactivity was an exponential decay process and immediately recognized its potential as a natural clock that could be used to determine the age of uranium-bearing rocks. But it was a precocious 18-year-old physics student at Imperial College, Arthur Holmes, who undertook the project of determining the first absolute geologic dates.[5] Starting in 1908 (the year after Lord Kelvin died), Holmes began seeking appropriate rock samples and separating minerals, especially zircon, that were known to contain uranium (U) but no lead (Pb) at the time of crystallization. He then needed to find the relative concentrations of uranium and lead in the mineral and used Rutherford's radioactive decay law, which quantified radioactivity as a function of time, to find how many years had elapsed since the mineral crystallized.[6]

The math is remarkably simple; the only numbers required are (1) the daughter: parent (Pb:U) ratio, which grows as a rock ages, and is independent of the (unknowable) original amount of parent material (see table 1); and (2) the decay constant for

TABLE 2.1. Simple illustration of the mathematics of radioactive decay

| | Example 1: Initial amount of parent = 100 | | | Example 2: Initial amount of parent = 32 | | |
|---|---|---|---|---|---|---|
| Time in half-lives since crystal formed | Amount of parent | Amount of daughter | Daughter: parent ratio | Amount of parent | Amount of daughter | Daughter: parent ratio |
| 0 | 100 | 0 | 0 | 32 | 0 | 0 |
| 1 | 50 | 50 | 1 | 16 | 16 | 1 |
| 2 | 25 | 75 | 3 | 8 | 24 | 3 |
| 3 | 12.5 | 87.5 | 7 | 4 | 28 | 7 |
| 4 | 6.25 | 93.75 | 15 | 2 | 30 | 15 |

*Note*: The daughter: parent ratio is the key to determining the age of a mineral and is independent of the original amount of parent material present.

the parent, which is essentially the probability that any given atom will decay in a certain amount of time—analogous to one's chances of winning the lottery in any year. The units of the decay constant are thus 1/time. Rutherford had estimated the decay constant for uranium from the number of radioactive emissions detected from a mass of uranium in a given time interval. The decay constant is inversely proportional to the more familiar idea of a half-life—the time it takes for half the parent material to decay to the daughter form. In other words, a small decay constant (low probability of a lottery win) means a long half-life (long wait to get rich), while a large decay constant means a short half-life (easy money!).

By 1911, in spite of the still-rudimentary understanding of the phenomenon of radioactivity and rather primitive laboratory facilities, Arthur Holmes had determined the absolute ages of a half-dozen igneous rocks whose relative ages on the fossil-based geologic timescale were bracketed by their relationships with sedimentary rocks. Three samples were from the fossiliferous Paleozoic and three from the murky, undifferentiated Precambrian. Even though some of the lead Holmes measured was not from the decay of the parent uranium but from another radioactive element, thorium, his dates are amazingly close to modern values (within tens of millions of years).

The very first rock analyzed, a granite from Norway thought to have formed in the Devonian Period (based on its cross-cutting relationships with fossil-rich sedimentary strata), yielded an approximate age of 370 million years—18 times longer than Kelvin's estimates of the age of the Earth. And a Precambrian metamorphic gneiss from Ceylon (Sri Lanka) was found to be 1.64 billion years old—two full orders of magnitude greater. Darwin's intuition was vindicated. Holmes would go on to become one of the preeminent geologists of the twentieth century. Kelvin's long-reigning proclamations became immediately irrelevant, because radioactivity not only provided a means of directly dating rocks but was also a source of internal heat that Kelvin had not incorporated into his calculations of the rate of planetary cooling. (Years later, Holmes would challenge another of Kelvin's fundamental assumptions, arguing that Earth cools mainly by convective, rather than conductive, heat loss). Most important, the geologic timescale could now be calibrated. Even the deepest reaches of geologic time could be fathomed; the Precambrian would no longer be an uncharted primordial wilderness.

### THE GENERAL SEDIMENT

In reality, it would take many more decades for the new science of geochronology (Earth time) to mature. The use of radioactive isotopes as high-precision geologic clocks required advances in nuclear physics, cosmochemistry (which concerns the stellar origins of the elements), petrology (study of igneous and metamorphic rocks), and mineralogy, as well as the development of new analytical instruments, particularly mass spectrometers capable of distinguishing among isotopes of a single element. There was also the nontrivial problem that geologic timescale so laboriously built by the Victorians, with fossils as timekeepers, was entirely based on sedimentary rocks. Any isotopic dates derived from these would reflect not the age of the sedimentary deposit but the time of crystallization of the igneous or metamorphic precursors from which its grains were derived. Assigning absolute ages to the fossil-based timescale has thus required finding serendipitous outcrops where sedimentary rocks of well-constrained *biostratigraphic* age happen to be interlayered with, or cut by, igneous rocks in such a way

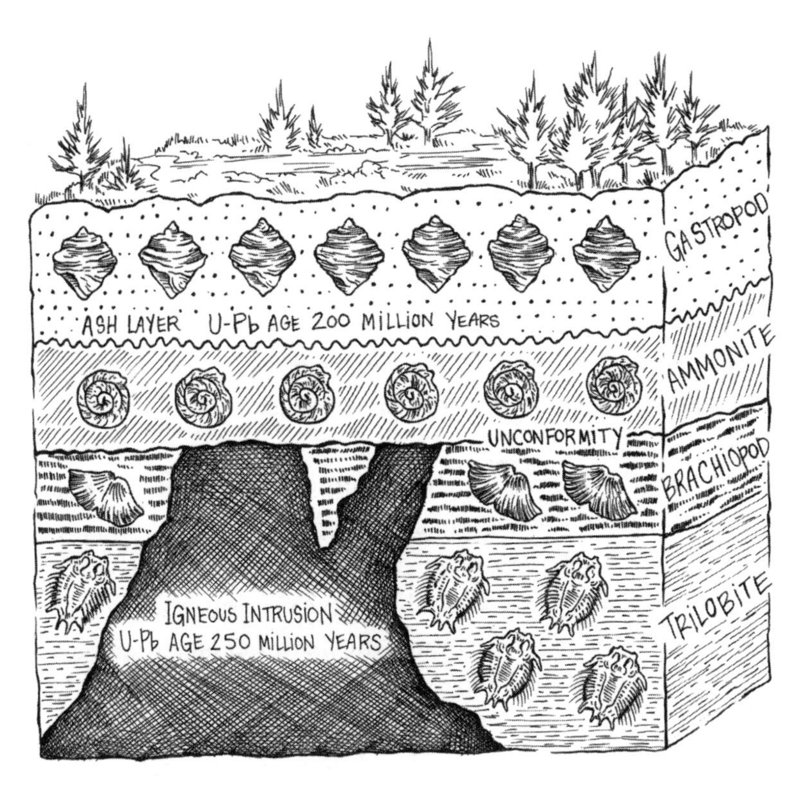


FIGURE 4. Cross-cutting relationships between igneous and sedimentary rocks allow calibration of the fossil-based timescale.

that allows an isotopic age to be tied directly with the fossil record (figure 4). Volcanic ash layers are ideal for this purpose, since they represent fresh igneous crystals that fell from the air in a geologic instant and were interleaved with the sedimentary and paleontologic archive of their day.

Ash layers within sedimentary strata reveal a subtle but fundamental idea about the way in which the rock record is written. Looking at layered rocks like the extraordinary sequence in the Grand Canyon, one tends to imagine that each stratum accumulated in the manner of a snowfall, blanketing a given area all at once, in a well-defined period of time. But this is not necessarily the right way to think about rock layers. Consider the beautiful white, almost pure-quartz St. Peter Sandstone of Ordovician age, exposed along river valleys in Minnesota, Iowa, Wisconsin, and northern Illinois. The St. Peter forms the picturesque hollow at Minnehaha Falls in Minneapolis and was for decades the source of silica for window glass made at the Ford plant in Saint Paul. During Prohibition, natural pockets in the St. Peter along the Mississippi River were enlarged into a network of caverns that housed speakeasies and secret warehouses beneath the Twin Cities.

The St. Peter Sandstone is crumbly, hardly even a proper rock, and when it falls apart into uniform, rounded grains in one's hand, it is easy to see that this is an ancient beach sand. But the St. Peter is found at the surface in four states, and is known from drilling to continue beneath Michigan, Indiana, and Ohio. No beach would cover such a vast area at any particular moment. Instead, the St. Peter records the gradual migration of beaches across the land surface as ancient shallow seas waxed and waned over millions of years. One day in the Ordovician, clouds of ash from a supervolcano eruption in the infant Appalachian Mountains hundreds of miles away fell out of the air over the midcontinent seas, leaving a thin, greenish clay layer across the region that is like a clearly dated diary entry. In some places, the ash occurs near the top of the St. Peter, but elsewhere, the sandstone lies far below this level, having been buried long before by other sediments at the time the volcano erupted. Thus, although the unmistakable St Peter sandstone is a continuous layer for hundreds of miles, it is not the same age everywhere. The more general idea is that except for layers that mark sudden regional or global events, like a great eruption or a meteorite impact, laterally extensive sedimentary units are not strictly *isochronous*—markers of the same moment in time. Instead, they record the slow march of depositional settings across the Earth's surface over time, as sea levels and environmental conditions changed. In geologic parlance, they are *diachronous*—that is, they transect time.

## THE TIME BUREAUCRATS

These days the geologic timescale is not merely a chart or even a multivolume treatise but a gigantic digital database that is administered by the formidable International Commission on Stratigraphy (ICS), the oldest and most important body within the International Union of Geological Sciences. The ICS maintains strict rules about how geologic units are named and defined, and it catalogues outcrops, rock formations, fossils, isotopic dates, geochemical data, and analytical protocols, in the never-ending task of mapping geologic time at higher and higher resolution.

Since the 1970s, the ICS has sought to identify specific sites around the world to serve as the international standards for the boundaries of each division of the geologic timescale. Such an outcrop is formally called a Global Boundary Stratigraphic Section and Point, or GSSP, but known colloquially among geologists as a "golden spike." These sites must have well-exposed rocks with biostratigraphically diagnostic fossils that straddle the boundary between the two time intervals, and they must be in places that can be protected from development or destruction. The location of the exact stratum representing the boundary at a given GSSP is often described in charmingly idiosyncratic detail. For example, the golden spike outcrop for the Cenomanian division of the Upper Cretaceous lies high in the French Alps and begins "36 meters below the top of the Marnes Bleues Formation on the south side of Mont Risou."[7]

The primary divisions of the geologic timescale—the eons, eras and periods—were largely defined by the work of British geologists in the nineteenth century, and the names of the Paleozoic Periods more strongly reflect that geographic influence: Cambrian, from the Latin name for Wales; Devonian for the county of cream teas; Carboniferous for the coal measures of northern England. But the finer subdivisions—the epochs and ages—reveal the subsequent, wholly international nature of the time-mapping project: the Jiangshanian and Guzhangian of the Cambrian; the Eifelian and Pragian of the Devonian; the Moscovian and Bashkirian in the Carboniferous. The ICS is like a temporal counterpart to the United Nations—a parliament of the past, whose jurisdiction is geologic time.

And the ICS, somewhat fussily, insists on maintaining the subtle but important distinction between time and the rock record of time. Geologic time is divided into eons, eras, periods, epochs and ages, and the corresponding rocks into eonothems, erathems, systems, series, and stages. Similarly, one should say "Early" or "Late" Ordovician (for example) when referring to time, but "Lower" and "Upper" when speaking of rocks. Time (*chronos*) could happen without rocks (representing *kairos*), but not the other way around. However, time vanishes, while rocks persist.

## PLUMBING THE DEPTHS OF TIME

Arthur Holmes's early efforts to obtain absolute ages from rocks, carried out before the structure of the atom and the existence of isotopes were even known, are analogous to Darwin's insights about heredity, which predated the discovery of genes and DNA. In both cases, it would take years before the rest of science developed the capacity to explore fully the implications of their visionary ideas. It was not until the 1930s that the complexity of lead isotope geochemistry was fully understood, one might say "plumbed." In 1929, Ernest Rutherford showed that there were two different parent isotopes of uranium, $^{238}U$ and $^{235}U$, which produced two different isotopes of lead ($^{206}Pb$ and $^{207}Pb$, respectively) at the end of long radioactive decay series with very different overall half-lives (4.47 billion and 710 million years, respectively). Soon after, Alfred Nier, a physicist at the University of Minnesota, identified another lead isotope, $^{204}Pb$, which was nonradiogenic—that is, lead that started as lead, and was not the product of radioactive decay. Nier had developed the essential instrument in isotope analysis, the mass spectrometer, which allows isotopes of a single element to be sifted out according to their atomic weight. And with the discovery of $^{204}Pb$, Nier recognized the potential application of these three lead species for dating rocks, and even the Earth.

Over geologic time, he realized, the abundances of $^{206}Pb$ and $^{207}Pb$ would have grown in a mathematically predictable way while the absolute amount of $^{204}Pb$ remained constant. In particular, the comparatively short half-life of $^{235}U$ would have caused Earth's inventory of $^{207}Pb$ to increase rapidly early in the planet's history, but then flatten off, like the cumulative earnings from a savings account with a high interest rate but from which rapid withdrawals are made. Meanwhile, the global stock of $^{206}Pb$ would have continued to accumulate from the slower decay of $^{238}U$—like the money earned at a lower interest rate in an account that is drawn down more slowly. (The unchanging amount of $^{204}Pb$ would be like money hidden under a mattress). In 1940, Nier and his students were about to put these ideas to the test using geologic samples. The work was interrupted when Nier—the son of German immigrants—was asked by Enrico Fermi to work on the Manhattan Project, which required the separation of the fissionable isotope $^{235}U$ from nonfissionable $^{238}U$.[8] Nier's spectrometer was the only instrument that could distinguish the two isotopes, and his lab was required to focus on the uncertain future rather than questions of the geologic past.

Immediately after the war, however, Nier set about measuring the Pb isotope ratios in deposits of galena (lead sulfide, PbS), the primary ore of lead, of different ages from around the world. Galena obviously has plenty of lead in it, but the lead doesn't take in uranium when it crystallizes. This means that lead isotope ratios in galena do not change over time and should instead reflect the particular mix of lead species that were available in the environment at the time the mineral formed. As Nier had predicted, the older samples had lower ratios of $^{207}Pb/^{204}Pb$ and $^{206}Pb/^{204}Pb$ (lead from "interest" vs. "mattress" lead). If the Earth had started with no $^{207}Pb$ or $^{206}Pb$, these ratios would be enough to determine the age the planet. But Nier knew that at the time of its formation, Earth had almost certainly inherited some "interest" lead from what had accumulated in the "bank accounts" of ancestor solar systems. Thus, determining the age of the Earth required knowing the primordial ratios of the various lead isotopes.

Nier also recognized a subtler problem: even a very ancient galena sample would not represent the primordial lead ratios for the Earth as a whole. Earth is not one uniform geochemical reservoir, like a planetary milkshake. Instead, it has unmixed itself over time. In its earliest days, the planet differentiated into a metallic core of iron and nickel and a rocky mantle that got most of everything else, including virtually all Earth's uranium. Ever since then, repeated partial melting of the mantle has generated the crust, which is much richer in uranium than either the bulk Earth or the mantle, in the way that butterfat is concentrated in the cream at the top of a bottle of raw milk. Nier's view was that although his lead isotope data broadly followed the expected pattern, some of the samples had probably had assimilated extra radiogenic lead ($^{206}Pb$ and $^{207}Pb$) derived from the decay of the "excess" uranium in crustal rocks and thus didn't exactly track the evolution of lead isotopes for the whole planet.

By the late 1940s, Arthur Holmes was a professor of geology at Edinburgh University, and had largely moved on to other major questions (such as the driving forces behind mountain building), but he had been following Nier's work and saw that it might allow the age of the Earth to be determined at last. He was especially intrigued by one specific sample Nier had analyzed, galena from a very ancient rock sequence in Greenland that had both low uranium concentrations and low lead isotope ratios. Holmes, always a big-picture, back-of-the-envelope thinker, was willing to make the assumption that meticulous Nier was not—that the Greenland galena provided something close to primordial whole-Earth lead isotope ratios. Conceptually, calculating the age of the Earth was simple: one just had to determine how much time it would take for the ratios to evolve from that primordial starting point to the values in younger galena deposits. In practice, however, the math was so difficult that Holmes had to purchase a mechanical computing machine to carry it out. After months of tedious calculations, Holmes published his minimum estimate for the age of the Earth: 3.35 billion years.[9] Geologists could finally relax into a luxurious abundance of time.

But now there was a new conflict between the timescales envisioned by geologists and physicists. According to the expansionary (Big Bang) theory of the Universe, which gained credence in the 1920s with Edwin Hubble's observation of galactic redshifts, the age of the Universe can be determined in a remarkably simple manner—almost trivial, in fact, compared with Holmes's lead isotope calculations for the age of the Earth. One just plots the velocity (distance/time) at which stars and galaxies are receding from Earth versus their distance from us. The slope of this line is called the *Hubble constant,* and the inverse of the slope, which has units of time, is the age of the Universe. In 1946, when Holmes declared Earth's age to be more than 3 billion years, the Universe was allegedly only 1.8 billion years old.[10]

## GEOCHEMISTS TAKE THE LEAD (OUT)

The embarrassing discrepancy between geologic and astronomical time remained unresolved for almost a decade, but as better estimates of stellar distances were made, and galaxies farther from Earth could be detected, the accepted value for the Hubble constant fell, and the age of the Universe grew. Then, in 1948, a young Iowa-born graduate student at the University of Chicago, Clair Patterson, struck upon a novel approach to the age of the Earth question. It was becoming clear that there were probably no surviving rocks that represent the original crust of the planet. Arthur Holmes had used the lead isotope ratios from the ancient Greenland galena as the closest available approximation to primordial values, but Patterson realized there was an even better source of information: extraterrestrial rocks—meteorites.

Meteorites represent preplanetary matter and fragments of ill-fated planets that formed at the same time as Earth and the rest of the Solar System. Unlike Earth rocks, which are in a constant state of modification and reincarnation through weathering, erosion, metamorphism, and melting, most meteorites have undergone no alteration in the vacuum of space since the formation of the Sun and planets. Any thin rind acquired from their passage through the atmosphere or time spent on Earth's surface can be pared off, revealing pristine material from the earliest days of the Solar System.

Patterson's approach was to use two different types of meteorites, with different compositions, to represent the original and modern values of lead isotopes in the Solar System, and then to repeat Holmes's strenuous calculations. Iron meteorites, which contain lead but no uranium, would provide the true primordial values. And stony meteorites, which contain both lead and uranium, would provide the modern bulk-Earth (well-mixed milkshake) value more reliably than any Earth rock could (see figure 5).

Once again, the idea seemed simple, but in practice required Herculean effort. Patterson found that he could not obtain consistent enough lead isotope results from duplicate samples to make meaningful age determinations. After systematically ruling out any flaws in his analytical methods, he realized what the problem was: there was so much ambient lead in the lab—on work surfaces, equipment, clothing, skin—that it was contaminating the meteorite samples before they could be analyzed. Over a period of nearly eight years, during which he moved to Caltech and then back to Illinois—this time to Argonne National Laboratory—Patterson developed the first "clean lab" (now an essential fixture in countless scientific and medical facilities) with a sophisticated air purification and ventilation system. In 1956, he finally obtained the number that remains the accepted age of the Earth: 4.55 billion ± 70 million years.[11] (*Requiesce in pace*, Darwin). After attaining, at age 31, the long-sought holy grail that had eluded

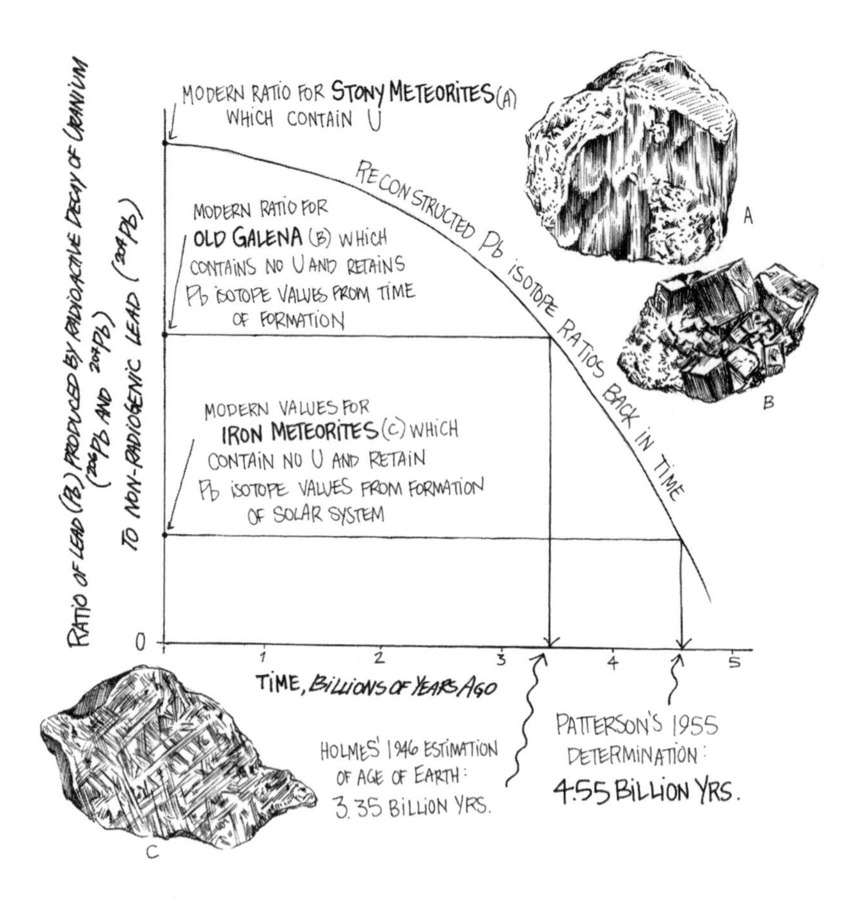


FIGURE 5. The logic behind using meteorites to date the Earth

geologists and physicists since the time of Hutton, Patterson left academe. He spent the rest of his life crusading to ban lead, already known then to be a neurotoxin, from paint, toys, tin cans, and gasoline. Reckoning the age of the Earth would seem to be a Nobel Prize–worthy accomplishment, but geologists aren't even in the running. Patterson did receive the prestigious Tyler Prize for Environmental Achievement just before his death, in 1995. Yet it seems understated recognition for a small-town Iowa boy who had stood up to giants: Kelvin, Hubble, and Big Oil.

## GEOCHRONOLOGY COMES OF AGE

Following the pioneering work of Nier, Holmes, Patterson, and others, the field of *geochronology*—the science of determining the age of geologic materials—expanded to include many other systems besides the uranium-lead decay series. Among the 92 naturally occurring elements, there are thousands of different isotopes, and most of these are radioactive (only 254 are stable). But not all radioactive isotopes are useful as geologic timekeepers. First, the half-life needs to be appropriate to the lengths of time being measured. Many isotopes have half-lives of days or seconds, and using them to measure geologic time would be like using a 12-inch ruler to measure the Alaska Highway. Also, because of the exponential nature of radioactivity, with half of the parent decaying away every half-life, there is almost no parent left after about 10 half-lives, no matter how much there was at the start (just as there is a limit to the number of times one can fold a piece of paper in half, no matter how big it is). Second, the parent isotope must be present in any rock or mineral being dated in high enough concentrations that it can be measured, and also yield measureable amounts of the daughter. The definition of "measureable" has changed over time, however; improvements in instrumentation now make it possible to detect elements that are in parts per billion and even parts per trillion concentrations in minerals.

Third, the daughter element should, ideally, not be incorporated into the mineral at the time of crystallization—the starting time for the isotopic clock—so that any daughter present in the sample is known to come from radioactive decay of the parent after the crystal became a closed system. This is a bit like the logic behind requiring students to use much-loathed "blue books" for exams; it ensures that they wrote all their answers for the test after they entered the classroom and the door was shut. (There are, however, mathematical techniques that can, in fact, correct for initial amounts of the daughter, just as an astute instructor might detect cheating on an exam.)

Finally, the daughter isotope should not be too prone to escaping from the mineral crystal, even though it is usually an ill-at-ease stranger in that setting. A parent atom, with its particular diameter and electric charge, will generally have had a comfortable place in the lattice of atoms in a mineral, bonding harmoniously with neighbors. But after the parent undergoes radioactive metamorphosis to a daughter isotope, it no longer fits in the crystal "chrysalis." It is a completely different element, with a different size and chemical proclivities. Given its discomfort in its parent's home, the daughter may try to leave the crystal, a possibility that becomes more likely if the rock is reheated sometime later in its history, and the framework of the crystal becomes more open to diffusion. Because the ratio of the daughter to parent isotope is the basis for determining the age of the sample (table 1), any loss of the daughter isotope will cause isotopic ages to be too young.

Because of these rather restrictive criteria, there only about a half-dozen parent-daughter isotope systems that can be used for dating rocks (table 2). These parent isotopes are a lasting legacy from the time of Earth's formation, inherited from precursor stars and solar systems, and some have absurdly long half-lives. The half-life of rubidium-87 ($^{87}Rb$), for example, at 49 billion years, is not only greater than the age of the Earth, but of the Universe (which is now thought, thanks to revised Hubble constant estimates, to be 14 billion years). This isn't an

TABLE 2.2. Parent-daughter isotope pairs most commonly used for geologic dating

| Parent isotope | Daughter isotope | Half-life (millions of years) | Parent isotope | Daughter isotope | Half-life (millions of years) |
|---|---|---|---|---|---|
| $^{238}U$ | $^{206}Pb$ | 4470 | $^{40}K$ | $^{40}Ar$ | 1280 |
| $^{235}U$ | $^{207}Pb$ | 710 | $^{147}Sm$ | $^{143}Nd$ | 106,000 |
| $^{232}Th$ | $^{208}Pb$ | 14,000 | $^{176}Lu$ | $^{176}Hf$ | 36,000 |
| $^{87}Rb$ | $^{87}Sr$ | 48,800 | $^{187}Re$ | $^{187}Os$ | 42,300 |

*Source*: Values from Faure, G., and Mensing, T., 2012. *Isotopes: Principles and Applications.* New York: Wiley.

inconsistency—it simply means that only a tenth of a $^{87}Rb$ half-life has elapsed since Earth formed, and that just a small fraction of the primordial $^{87}Rb$ has so far decayed to strontium-87 ($^{87}Sr$). But because rubidium is a common trace element in many minerals, both $^{87}Rb$ and $^{87}Sr$ occur in high enough concentrations to be measureable and geologically useful.

Some rocks, like granite, have two or more minerals that can each be dated using a different parent-daughter isotope system, and it is common to find that these minerals yield different ages. This is another geologic observation that has been seized upon by Young-Earthers in an attempt to "debunk" the geologic timescale, but it would in fact be surprising if all the minerals in an igneous rock such as granite, formed when a mass of magma cools slowly deep underground, did report the same isotopic ages. The reason is that the *closure temperature* for each mineral—the point at which the crystal "doors" become shut to diffusion—is different for each parent element in each mineral species. Knowing these specific closure temperatures allows the cooling history of subsurface magma bodies—called *plutons*, for Pluto, Roman god of the Underworld—to be reconstructed in great detail. For example, combined U-Pb, Rb-Sr, and K-Ar dating of minerals from the Tuolumne granites in Yosemite National Park reveals that they remained above 660°F for more than 3 million years.[12] The granites that now form the sublime peaks of the High Sierra represent magma chambers that fed mighty Jurassic volcanoes, long since eroded away. Understanding how long a magmatic plumbing system may remain active is relevant to predictions about the risk of eruptions in places like Yellowstone, where mud pots and geysers hint at unrest in the Underworld.

## RADIOCARBON DAYS

The best-known isotope used for dating, carbon-14 ($^{14}C$), is in many ways an oddball and differs from other parent isotopes in several important ways. With an extremely short half-life of just 5730 years, it can't be used to date anything older than about 60,000 years (so its use in geology is limited), and it doesn't represent a primordial species—after 4.5 billion years, it would no longer exist. Instead, it is a *cosmogenic* isotope that is continuously regenerated in Earth's uppermost atmosphere by cosmic rays—high-energy radiation from space. Cosmic rays are thought to come mainly from distant supernova events, in which old stars explode in a spectacular final extravaganza (producing new elements and isotopes that may be incorporated into future planetary systems). Because of concern over long-term exposure to cosmic rays, pilots and flight attendants are typically limited to a certain number of long-haul high-altitude flights each year.

Carbon-14 is produced when a nitrogen-14 ($^{14}N$) atom high in the atmosphere is struck by a cosmic ray with enough energy to knock a proton out of the nitrogen nucleus. Some of this

$^{14}C$ makes its way to the surface of the Earth and is taken in by photosynthesizers (algae, plants) and the organisms that consume them (fish, fungi, sheep, people). As long as a plant or animal lives, photosynthesizes, breathes, and/or eats, its blend of carbon isotopes (stable $^{12}C$ and $^{13}C$, as well as radioactive $^{14}C$) will reflect the relative abundances of carbon in the environment. But when the organism dies, its carbon inventory becomes fixed, and the radioactive $^{14}C$ gradually ticks away while the stable carbon isotopes remain. In contrast with other isotopic dating methods, in which the daughter/parent ratio is used to determine the age of a sample, $^{14}C$ ages are based on the *activity* of the carbon present—the number of decay events per unit time per gram of carbon. The reason is that $^{14}C$ decays back to $^{14}N$, a gas that will tend not to be retained in the sample.

Carbon-14 dating is an important tool in archeological and historical research and can be used to date a wide variety of materials containing biogenic carbon: wood, bone, ivory, seeds, shells, linen, cotton, paper, peat, and the like. Even ocean water can be dated, because it has a small amount of carbon dioxide dissolved in it. Some deep-ocean water from the North Pacific yields $^{14}C$ ages of 1500 years[13]—meaning those waters have not interacted with the atmosphere since before the birth of the prophet Muhammad.

But the uncertainties in $^{14}C$ ages are proportionally rather large compared with geologic age determinations because the rate of production of $^{14}C$ in the upper atmosphere has varied over time owing, among other things, to fluctuations in Earth's magnetic field, which partly shields the planet from cosmic ray bombardment. Carbon-14 dates can be corrected using tree rings, those low-tech but reliable timekeepers, because only the outer part of a tree is actively exchanging carbon with the environment in a given year, and so each ring will have a different $^{14}C$ age. By correlating the oldest rings in living trees with the youngest rings in ancient wood preserved in bogs and archeological sites, the tree ring record can be extended back more than 10,000 years, and $^{14}C$ ages can be adjusted accordingly. Growth bands in corals (made of calcite, $CaCO_3$) provide a somewhat lower-fidelity record than tree rings but make it possible to calibrate $^{14}C$ ages still further back in time. Nevertheless, the uncertainties for $^{14}C$ dates are large—on the order of hundreds to thousands of years (5% to 10% of an object's actual age).

Humans have further complicated radiocarbon dating in two ways. First, aboveground nuclear tests in the early Cold War days injected large amounts of $^{14}C$ into the atmosphere, which must be corrected for in very recent samples. This is why $^{14}C$ ages are typically reported as years before 1950. Second, a century of burning fossil fuels with "dead" carbon has shifted the mix of isotope values in the atmosphere. This is called the *Suess effect,* for the Austrian physicist Hans Suess, who first recognized it in 1955[14] (and who had been working for the German nuclear program at the time of the Manhattan Project in the United States). While the Cold War's $^{14}C$ will slowly dissipate, the Suess effect continues to grow.

## PRODIGAL DAUGHTERS

As mass spectrometers became more accessible to the academic masses in the late 1950s and '60s, geochronology came into its own as a new subdiscipline with dedicated faculty lines and graduate programs. Among the first isotope systems to be widely used for geologic dating was the potassium-40–argon-40 ($^{40}K$ -$^{40}Ar$) parent-daughter pair, because potassium is very abundant in many igneous and metamorphic rocks, and even lower-precision instruments could detect both parent and daughter. The original K-Ar method is still perfectly good for young rocks with simple thermal histories. It remains an important tool, for example, in determining the ages of sedimentary deposits containing fossils of human ancestors like "Lucy" that are conveniently interbedded with volcanic ash layers in the magmatically active East African Rift Valley.

A problem with the K-Ar system is that the daughter is an apple that falls far from the parental tree. Potassium is a large, sociable ion ready to offer an electron to other elements, while argon is a compact, self-contained noble gas with completely filled electron shells and no tendency to bond with anything. So given any chance—a position at the edge of a crystal that allows an easy exit, a crack that offers a shortcut, a metamorphic heating event that opens the crystal's doors to diffusion—daughter argon atoms will leak out. The calculated age for the host mineral will then be younger than the true geologic age, but there is no way of knowing by how much. The plus or minus value will reflect the analytical uncertainty arising from the limits of the laboratory instruments, not the actual imprecision of the date.

The limitations of K-Ar dating began to be especially clear in the 1960s when the method was applied to old rocks from the Canadian Shield, which had long, multistage histories of deformation and metamorphism. Age determinations were sometimes inconsistent with field evidence for the relative ages of rocks. In some cases, so much argon had seeped out of minerals deep in the subsurface that it lingered in adjacent rocks, leading to cases in which the K-Ar age determinations were actually too old. Young-Earth creationists still seize on these ambiguities and suggest that the whole enterprise of geochronology is hopelessly flawed. But by the 1970s, geochronologists had developed a powerful variation on K-Ar dating that yields both higher-precision ages and provides information about whether argon loss (or gain) has occurred.

In the new technique, a potassium-bearing sample is bombarded with neutrons, and this converts the $^{40}K$ in the specimen to a short-lived isotope of argon, $^{39}Ar$, which then acts as a proxy for the parent. The sample is next heated slowly in what amounts to the laboratory equivalent of a metamorphic event. Both types of argon—$^{39}Ar$, representing the parent, and $^{40}Ar$, the daughter produced by radioactive decay—begin to leak out. As the temperature is increased incrementally, the crystal begins to exhale more argon, which is captured and analyzed in batches. The $^{40}Ar/^{39}Ar$ ratio (really the daughter/parent ratio) is used to determine an apparent age of the sample at each step. Typically, the ages obtained from the first few samples of captured argon—representing the outside of the crystal, where geologic argon escape would have been easiest—are younger than those for the interior. If the apparent ages obtained with continued heating stabilize around a consistent value—what geochronologists call an "$^{40}Ar/^{39}Ar$ plateau age"—then there is good reason to conclude that the interior of the crystal has not experienced significant argon loss and that the age is geologically meaningful.

## DATES WITH DESTINY

Probably the most famous application of the argon-argon dating method was the conclusive identification of the crater formed by the meteorite impact that doomed the dinosaurs at the end of the Cretaceous Period. The meteorite hypothesis for the dinosaur extinction was first proposed in 1980 by the father-son team of Luis Alvarez, a Nobel Prize–winning physicist, and his son Walter Alvarez, a geologist at Berkeley. Walter had been working in the central Italian Apennines, where the crinkling of the crust into recently formed mountains has raised a sequence of late Mesozoic and early Cenozoic marine limestones above sea level.[15] One of these, the Scaglia Rossa (essentially, "red rock")—a beautiful pink limestone that defines the color palette of many Italian houses, castles, and cathedrals—contains an uninterrupted chronicle of ocean conditions before, during, and after the Cretaceous extinction. There are no dinosaur bones in the Scaglia Rossa, since it accumulated on the seafloor on the continental shelf of Africa, but the extinction event is clearly recorded by an abrupt change in the nature and number of microscopic fossils and by a distinctive half-inch-thick dark-red layer of clay.

Walter Alvarez wondered how much time this clay layer—a mute witness to global apocalypse—represented. His father Luis, another Manhattan Project alumnus, had access to an instrument at Lawrence Berkeley Laboratory that could detect trace elements in materials at the parts per billion (ppb) level. He suggested measuring the boundary clay's concentration of certain rare metals in the platinum group, such as iridium, that are delivered to earth's surface mainly by a slow but constant rain of micrometeoritic dust (you can even collect micrometeorites, many of which are magnetic, from your roof over a period of months[16]). The average rate of this metallic "rainfall" over the past 700,000 years is known from Antarctic ice cores, and assuming it was about the same in the Cretaceous, measuring the metal content of the boundary clay would allow an estimate of how long it had taken that layer to accumulate. The logic was essentially the same as that used by Victorian geologists who attempted to rebut Kelvin: sum up the total amount of accumulated stuff (sediment, or iridium) and divide by the best estimate of its rate of accumulation to estimate elapsed time.

To have some idea of background concentrations of iridium, the Alvarezes analyzed closely spaced samples not only from the clay layer but also from the limestone below and above the boundary. They found that the concentration of iridium went from about 0.1 ppb in the underlying limestone to more than 6 ppb in the clay. The absolute amount seems small, but the anomalous spike—a 60-fold increase—was dramatic. It meant either (1) the clay layer represented a very long period of time during which meteoritic dust rained down slowly, yet very little normal sediment accumulated; or (2) a very large amount of meteoritic material had been delivered all at once to Earth by an object on the order of 10 km (6 mi) in diameter. Neither of these seemed likely, but of the two, the second seemed less unlikely.

However, this deus ex machina explanation ran counter to the deeply instilled habit of uniformitarian thinking in geology and its Lyellian aversion to invoking catastrophic causes. Also, the seemingly thin thread of evidence—a tiny increase in a strange element in a thin clay layer—was not convincing to many paleontologists who had spent their lives studying the fossil record for clues to the Cretaceous extinction. But as similar iridium anomalies were documented at other sites around the world where uppermost Cretaceous rocks are exposed, the story gained momentum. The new question became, Where was the crater?

By the late 1980s, a trail of tektites—spheres and teardrops of glass formed from the melting of rock in high-energy impacts—pointed to the Caribbean region as the most likely location of the end-Cretaceous ground zero. But it wasn't until 1991, more than 10 years after the original meteorite impact hypothesis was proposed, that a crater of the right approximate age and size was identified—a 190-km (120-mi)-wide depression largely buried by younger sediment off the north coast of Mexico's Yucatan Peninsula. It was named the Chicxulub crater after the nearest seaside village. The following year, the publication of argon-argon ages of in situ melt glass from drill cores taken at the center of the crater was enough to change the minds of geologists who were still skeptical about whether this was the site of the cataclysm. The weighted mean of the $^{40}Ar/^{39}Ar$ plateau ages for three samples was 65.07 ± 0.10 million years—the International Commission on Stratigraphy's precise chronometric definition of the end of the Cretaceous Period.[17]

## PARSING THE PRECAMBRIAN

In the context of Earth's history, dinosaurs are like attention-hogging celebrities who get a disproportionate share of media coverage when there are so many other important stories to be followed. While I respect all rocks, I must confess to some prejudices. Having grown up on the edge of the Canadian Shield—the old core of the North American continent—I have a deeply instilled predilection for rocks with at least a billion years behind them. Like wine and cheese, rocks grow more interesting as they age, accumulating flavor and character. For one thing, most Precambrian rocks have survived long enough to have been caught up in at least one episode of tectonic upheaval and carried to depths far from their native habitats, then against all odds, to have found their way back to the surface. Young rocks communicate in plain prose, which makes them easy to read, but they typically have only one thing to talk about. The oldest rocks tend to be more allusive, even cryptic, speaking in metamorphic metaphor. With patience and close listening, however, they can be understood, and they generally have more profound truths to share about endurance and resilience.

Even before Claire Patterson's decisive determination of the age of the Earth, isotopic dates from Precambrian rocks were revealing how greatly the Victorian fossil-based timescale had distorted geologists' perceptions of geologic time. Rocks of lowermost Cambrian age were known to be about 550 million years old, but rocks in the Canadian Shield were yielding ages greater than 2 billion years. And a 4.5 billion-year-old Earth meant that the quasi-mystical Precambrian, once viewed as the irretrievable infancy of the Earth, actually includes its childhood, youth, and most of its adulthood—eight-ninths of its total age. Even today, there is a lingering habit of overemphasizing the Phanerozoic—the eon of "visible life," from the Cambrian to today. Most textbooks of historical geology still devote only a perfunctory chapter or two to the Precambrian and then move quickly on to the "real" story. Little by little, high-resolution geochronology, and in particular a new generation of uranium-lead analyses, is correcting this persistent temporal bias.

Just as people have no memory of their birth or first year of life, Earth has no direct record of its formation or earliest days. Earth's own chronicle of its past begins with faint, cryptic entries from between 4.4 to 4.2 billion years ago, in the form of a few tiny crystals of the durable mineral zircon that were preserved as grains in an ancient sandstone in the remote Jack Hills of western Australia. The significance of these oldest of all Earth objects has been hotly contested since the announcement of their discovery in a now-famous paper in *Nature* in 2001.[18]

Zircon is a geochronologist's dream (and was the mineral Arthur Holmes used in the very first geologic age determinations). It accepts uranium but not lead into its structure at the time it crystallizes. And because uranium has two radioactive parent isotopes that decay to different lead daughters, there is a built-in cross-check for whether any daughter has been lost.

If the $^{206}Pb/^{238}U$ and $^{207}Pb/^{235}U$ ages match, the dates are said to be *concordant*, and this is good evidence that no lead loss has occurred. The precisions of concordant U-Pb zircon dates are astonishing: the oldest Jack Hills zircon gave an age of 4404 ± 8 million years, or an uncertainty of only 0.1%—far better, proportionally, than $^{14}C$ dates. Still, all is not lost even if lead was lost; statistical analysis of a collection of *discordant* zircons from a given rock can yield not only their crystallization age but often the age of the metamorphic event that led to the lead loss.

In addition, zircon is a physically tough mineral, capable of withstanding abrasion and corrosion that others cannot, and it has a very high melting temperature, so it can come through metamorphism without losing its "memory" of its earlier days. As geochronologists are fond of saying, "zircons are forever" (in contrast with diamonds, which, as high-pressure mantle-derived minerals slowly but inexorably revert to graphite at Earth's surface). Old zircon crystals commonly have concentric zones that are almost like tree rings—the core of the crystal records its original crystallization from a magma, and the successive rings reflect growth during later metamorphic events (figure 6). The most advanced generation of mass spectrometers—the Super High Resolution Ion Microprobe, or SHRIMP—can find isotope ratios for individual "growth rings" as narrow as 10 microns, less than the width of a hair. The extremely old ages obtained for the Jack Hills zircons came from the interiors of crystals with complex overgrowths. Just as the rings of one old tree may contain a climate record for a whole region, a single ancient zoned zircon crystal can chronicle the tectonic history of a continent.

The astounding age of the Jack Hills zircon grains is even more surprising in light of the fact that zircon forms almost exclusively during the crystallization of granites and similar

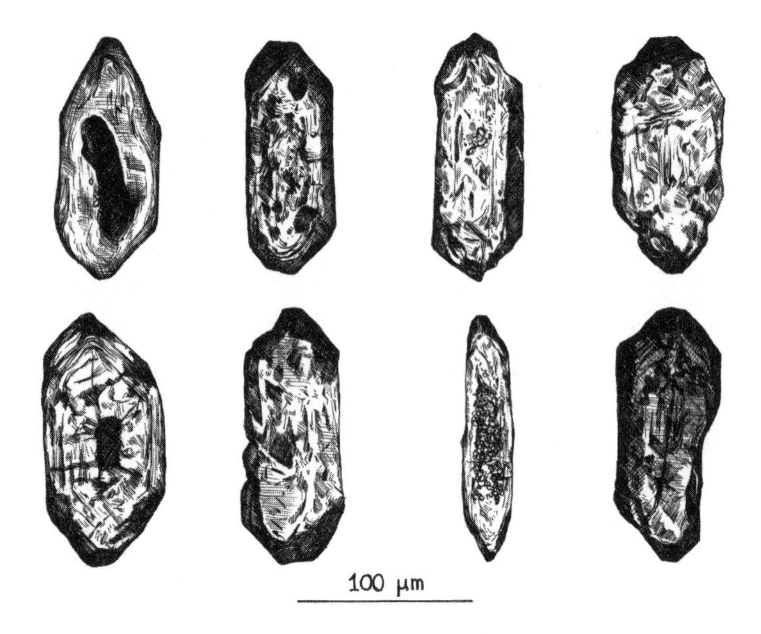


FIGURE 6. Zircon crystals with growth bands

igneous rocks, which are the foundations of the continents. Granites represent "evolved" magmas, meaning they are difficult to form in just one stage of melting from the Earth's mantle (the ultimate source of all crustal rocks). Today, granitic rocks come mainly from the forges of subduction-zone volcanoes like Mount Rainier and are derived by partial melting of preexisting crust, usually in the presence of water (more about this in chapter 3). So if the Jack Hills zircons were forged in this modern way, their existence suggests the dizzying prospect of a still-earlier crust that had formed, cooled, and then remelted within the first 150 million years of the planet's origin. Equally surprising, the ratios of different oxygen isotopes in the old zircons suggest that the magma from which they crystallized had interacted with relatively cool surface water.

Abandoning customary scientific restraint in their concluding paragraph, the authors of the 2001 *Nature* paper boldly suggested—on the basis of a few crystals smaller than fleas—that not only continents and oceans existed on Earth 4.4 billion years ago but that if surface water was around, perhaps there was even life.

## THINKING LIKE A PLANET

The Jack Hills zircon paper, one of the most cited in all the geologic literature, was a virtuosic culmination of a century of isotope geochemistry and required the most advanced analytical methods available. Yet, in its audacious inductive speculations and strong predilection for uniformitarianism, it is remarkably similar to the very first work of modern geology: Hutton's *Theory of the Earth*. Whether, in fact, the early Earth should be viewed through strictly uniformitarian spectacles is currently a matter of lively debate among geologists. There are compelling reasons to suggest that Earth's habits were different in its first 2 billion years.

But the story of how the still-unfinished Atlas of Deep Time has evolved, from Siccar Point to Chicxulub to Jack Hills, makes it clear that mapping time has been a very human endeavor that requires just this kind of give-and-take. It has involved a great variety of minds—visionary thinkers like Hutton and Lyell not too obsessed with details; attentive fossil-hunters like William Smith who are; polymaths like Darwin and Holmes who see connections across disciplines; fastidious instrumentalists like Nier and Patterson; bureaucracies like the International Commission on Stratigraphy; and legions of hardy, anonymous field mappers (including a few jocks) who understand both *chronos* and *kairos*, and how to turn rocks into verbs.

## CHAPTER 3

# THE PACE OF THE EARTH

How many years can a mountain exist
before it is washed to the sea?

—BOB DYLAN, 1963

### EPHEMERAL GEOGRAPHIES

One of my earliest school memories is of watching a film about the emergence of Surtsey, a volcanic island off the coast of Iceland that began to rise from the Atlantic late in 1963. The black-and-white footage showed explosive spires of steam and ash creating a blank new world of coal-dark cinder not yet on any map. A ship captain had been the first to notice the eruption and had initially thought it was another vessel on fire. To my impressionable young mind, the idea of new land forming was thrilling; it suggested a secret life-force inside the impassive, stony-faced Earth. Between 1963 and 1967, Surtsey built itself up from a submarine ridge 130 m (425 ft) below sea level to a small cone more than 170 m (570 ft) above it. At its maximum extent, Surtsey had an area of about a square mile. But after the eruptions ceased, its destruction by erosion, settling, and subsidence was nearly as rapid. Today, it has been reduced to about half its 1967 size and is expected to disappear entirely by 2100 (or sooner, depending on the rate of sea level rise). To my still impressionable middle-aged mind, it is somehow unsettling to have seen the arc of Surtsey's life—the birth, youth, brief prime, and inexorable demise of a landmass.

To Hutton, Lyell, and Darwin, most geologic processes seemed imperceptibly slow, and for decades, geologists drummed this idea into the public consciousness. But today, thanks to high-precision geochronology, direct satellite observation of Earth processes from space, and a century of monitoring the planet's vital signs—temperature, precipitation, river flow, glacier behavior, groundwater reserves, sea level, seismic activity—many of the geologic processes that once seemed beyond the reach of direct human observation can now be clocked in real time. And we are finding that the pace of the planet is neither as slow nor as constant as was previously thought.

## BASALT OF THE EARTH

Hutton's original epiphany that the age of the earth was effectively infinite compared with human lifespans arose from his recognition that the unconformity at Siccar Point represented the time needed for a mountain belt to form and be beveled again to a flat plain. So how long, exactly, would this take? The forces behind mountain building were not known until about 175 years after Hutton's death—in fact, around the time of Surtsey's birth in the 1960s, when plate tectonic theory finally explained how the solid Earth works. Today we realize that the tempo of mountain growth is ultimately set by the formation and destruction of ocean basins.

Unlike the continents, which are a messy amalgam of many different rock types of a wide range of ages and individual histories, the ocean crust is simple and homogeneous. It's all basalt—the black volcanic rock of Surtsey—and it's all produced in the same way: by partial melting of the Earth's mantle beneath submarine volcanic rifts, marked by high-standing midocean ridges. Counter to fanciful depictions in fiction and film, the mantle (which constitutes more than 80% of Earth's volume) is not a vat of seething magma but solid rock—though it flows over geologic timescales. Every few hundred million years, the mantle overturns itself in the manner of a gargantuan lava lamp, through the process of thermal convection: hotter, buoyant rock from depth rises while cooler, denser rock sinks. Mantle convection is Earth's main heat-loss mechanism (contrary to Lord Kelvin's erroneous assumption that the mantle was static and Earth had cooled over its lifespan through conduction). Arthur Holmes was among the first to suggest, in the 1930s, that the mantle convects; today, high-pressure experiments simulating the behavior of minerals at mantle depths show that convection of rock in Earth's interior is inevitable.

Midocean ridges are thought to coincide with areas of convective upwelling, where the Earth's crust is forced to stretch and thin above the rising plume of hot rock. Paradoxically, however, no melt forms until the ascending rock has lost much of it heat. So what makes still-solid mantle rock melt as it nears the surface? The mechanism is counterintuitive—not an input of heat but a decrease in pressure. Unlike water, a completely abnormal compound from which most of us derive our understanding of phase changes, rock behaves the way a proper substance should: it expands on melting and contracts on freezing. This means that if a rock is close to its melting temperature at some depth in the Earth and is depressurized (e.g., by rising closer to the surface), the lower-density phase—melt—becomes favored, and magma is formed. This phenomenon is called *decompression melting* and can happen even if the rock is actually cooling, as long the pressure is decreasing faster than the temperature is. (Decompression melting is especially hard for skiers and skaters to understand, since the opposite behavior by water—melting under *higher* pressure—is the very basis for winter sports that involve slippery surfaces).

On Earth today, after 4.5 billion years of cooling through mantle convection, upwelling mantle rock does not carry enough thermal clout to undergo wholesale melting. Instead, magmas at ocean ridges represent the components in mantle rock that melt at the lowest temperature. This partial, or fractional, melting is what generates basalt, which has a different composition—higher in silica, aluminum, and calcium, and lower in magnesium—than its parent, the mantle.

As each new pulse of basaltic melt rises and fills the central axis of an oceanic rift, the previous batches, now frozen into rock again, are displaced symmetrically outward in the process called seafloor spreading (see figure 7). The most recently erupted basalt is warmer and less dense than the slightly older rock it has pushed aside, and each generation cools progressively as it moves away from its birthplace at the rift. This is the reason that the midocean ridges stand high, like a soufflé fresh from the oven. In fact, one of the clues that led to the epiphany of plate tectonic theory in the early 1960s, when deep-seafloor maps first became available, was that the cross-sectional form of the ocean ridges is essentially a pair of mirror-image cooling curves—the shape of two skis placed on the floor tip to tip.

## ALL OVER THE MAP

Let us pause to contemplate how incredible it is that most of the Earth's surface—the deep-ocean floor—had not been mapped until the middle of the twentieth century. Even today, the topography of much of the seafloor is known to a resolution of only about 3 miles; bathymetric charts of the ocean are about

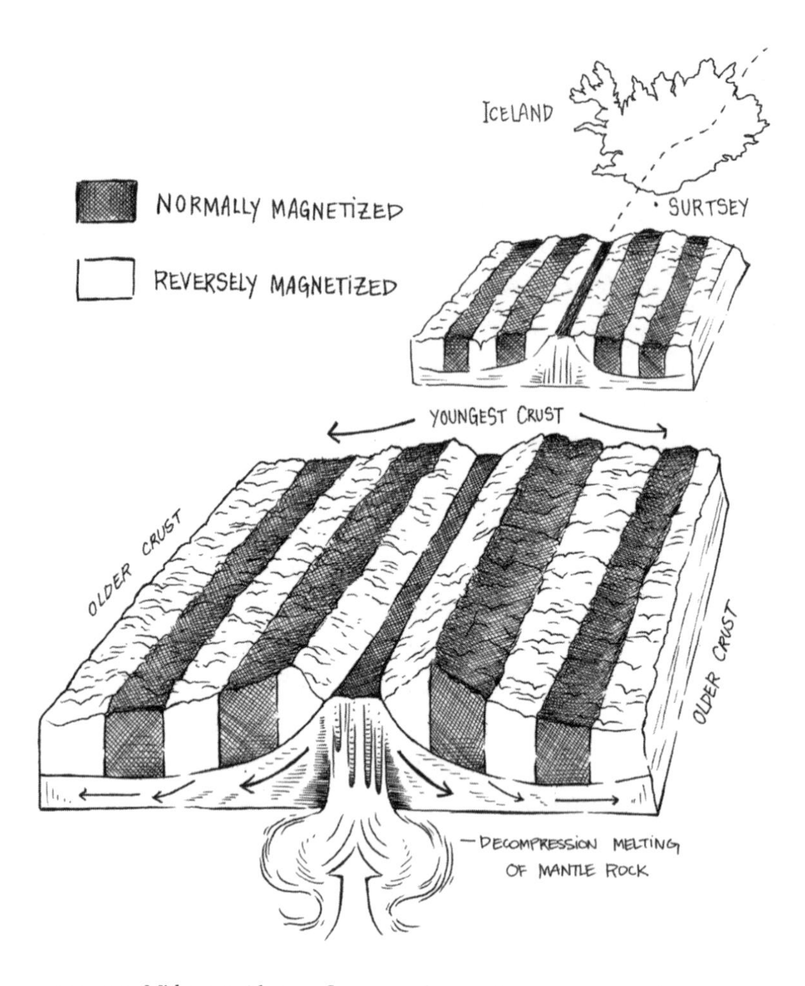


FIGURE 7. Midocean ridge, seafloor spreading, and magnetic reversals

100 times "blurrier" than current maps of the surface of Venus and Mars.[1] Still more incredible is the fact that one person almost single-handedly created the first maps of two-thirds of the planet yet is unknown to the average citizen of Earth (while Amerigo Vespucci, whose cartographic credentials are suspect, has two continents named for him). The unsung mapmaker was

Marie Tharp, who earned a master's degree in geology from the University of Michigan, worked briefly for an oil company, and then in 1948 became a drafter for a new oceanographic project led by Maurice Ewing at Columbia University.[2] For years, Ewing's all-male team of graduate students collected sonar soundings of the ocean floor while Tharp laboriously transformed the linear strings of depth readings into three-dimensional topography.

Tharp's exquisite shaded relief maps, painstakingly drawn in pen and ink, revealed that the seabed—previously thought flat and featureless—had a rugged, globe-encircling range of ridges and terrifyingly deep trenches. By 1953, she had noticed that the high ridges all had central down-dropped valleys and speculated that this might be evidence for crustal stretching. She shared her idea with another member of Ewing's group, Bruce Heezen, who infamously dismissed it as "girl talk." But Heezen and Tharp became close collaborators at Columbia, producing a series of seafloor maps that revolutionized geologists' view of the Earth. In 1963, when two British geologists first articulated the concept of seafloor spreading in a paper in *Nature*[3] (and Surtsey was demonstrating the process), Heezen—and much later, the rest of the geologic community—acknowledged that Tharp had been right.

The authors of the 1963 paper, Fredrick Vine and Drummond Matthews, proposed seafloor spreading on the basis of a perceptive geometric argument rather than firsthand geologic observation (the ridges would not be directly seen or sampled for another decade). Vine and Matthews had access not only to Tharp's maps but also to data from the U.S. and Royal Navies on the magnetic signatures of rocks at the bottom of the ocean. They noted that both the ridge topography and the magnetic intensity readings had mirror symmetry moving outward from the ridge crest—that is, bands of similarly magnetized rocks ran in parallel stripes on either side of the ridge (see figure 7). The ridge heights fell off with distance in just the way one would expect for deflating soufflés or cooling and contracting rocks. The symmetrical pattern of magnetic stripes suggested that successive generations of ocean crust had formed at the ridge, cooled enough for their iron-bearing minerals to align with the ambient magnetic field, and then been cleaved in half and displaced outward in a great conveyor system. Meanwhile, the polarity of Earth's magnetic field had repeatedly reversed itself, the north and south geomagnetic poles switching places on an erratic schedule (a second revolutionary inference in a paper that is barely three pages long).

By the early 1970s, age determinations for seafloor samples from deep-ocean drilling, as well as correlation of the marine magnetic record with magnetic reversals in well-dated volcanic sequences on land, had created a new way to demarcate geologic time, and the *geomagnetic* timescale was grafted onto the biostratigraphic (fossil-based) and *geochronologic* (radioisotope-calibrated) timescales. Today, with the date of each magnetic field reversal well constrained, it is possible to determine the age of a rock anywhere on the seafloor without even getting a physical sample—simply by counting how many magnetic stripes it is away from the ridge.

On a map showing the ages of seafloor in the world's oceans, the most striking pattern is that the swaths of rock of any given age are much wider in the Pacific Ocean than in the Atlantic. Since the start of the Cenozoic Era 65 million years ago (i.e., since the demise of the dinosaurs), seafloor spreading rates in the Atlantic have averaged about 1 cm (ca. 1/2 inch) per year, which is on the order of the rate at which one's fingernails grow. It's fast enough that at Thingvellir in Iceland, one of the few places where an ocean ridge stands above sea level—and the site the Vikings chose in AD 930 for their annual parliament meeting, the Althing—the visitor center was built to be as wide as the amount by which the crust has stretched since Viking times.

On the other hand, the rate of spreading in the Atlantic is slow enough that a species of green sea turtle (*Chelonia mydas*) from Brazil that has made an annual swim to a high spot on the Mid-Atlantic Ridge to breed and nest since the time of the dinosaurs hasn't seemed to notice that the ridge is now nearly 1100 km (700 mi) farther distant. Luckily the turtle's natal beaches weren't in the Pacific, where spreading rates are almost an order of magnitude faster, at close to 10 cm (4 in.) per year (a little slower than the "velocity" of hair growth). If these rates simply reflected the pace of mantle convection, why would that pace be more vigorous beneath one ocean than another?

## PLATES PULL THEIR WEIGHT

Marie Tharp's marvelous maps hold clues to the disparity in rates of plate motion in the two oceans. In particular, they show profound differences between the edges of the Pacific and Atlantic basins: the margins of the Atlantic Ocean are mainly shallow continental shelves, like the area off the coast of the eastern United States, where water depths are less than about 200 m (660 ft), and submerged crust gives way gradually to emergent land. The margins of the Pacific Ocean, in contrast, are delineated by vertiginous chasms, like the one off the west coast of South America, whose deepest points lie more than 8000 m (24,000 ft) below sea level. These trenches mark the sites of subduction, where old, cold, ocean crust—with the same instinct as the Brazilian turtles—returns to its place of origin.

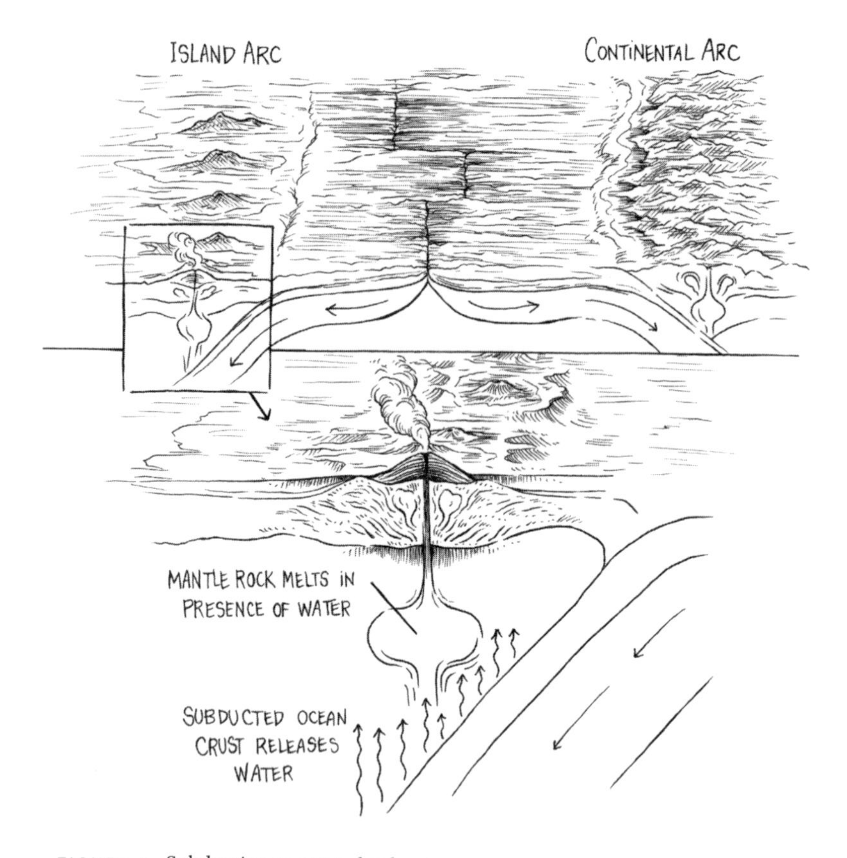


FIGURE 8. Subduction zones and volcanic arcs

When sea floor basalt is around 150 million years old, and hundreds of miles from its natal ridge, it has become about as dense as the underlying mantle and sinks back into Earth's interior at a slant, pulling the rest of the plate behind it, like a blanket sliding off a bed (figure 8). This "slab pull" force is almost certainly what sets the tempo for the rapid seafloor spreading in the Pacific—its rifts are simply keeping up with rates of subduction on its edges. In contrast, the Atlantic spreading rate probably reflects something close to the mantle's natural, stately pace. Earth's convective behavior should therefore be considered an "active lid" system, in that the plates do not merely dance to the mantle metronome but in some cases set their own meter, and this ultimately dictates how fast mountains grow. To build mountains, however, we first need to cook up some continental crust, and the recipe takes us back to the midocean ridges.

## WATER WORKS

Vine and Matthews correctly interpreted the morphology of ocean ridges as a record of the cooling of successive batches of basalt. But fresh ocean basalt does not give up its heat passively, like our soufflé cooling quietly in the kitchen. Instead, heat is robbed from it by frigid ocean water, which streams through cracks and pores, jealously steals joules, then makes a high-speed getaway through chimneylike underwater geysers called *black smokers*. The water also pilfers elements like calcium from the young rocks, and leaves behind sodium, mediating the salinity of the oceans. (This was unknown to John Joly when he tried to estimate the age of the Earth based on the salinity of the sea. His value of 100 million years was not meaningless—but it represented the typical residence time of sodium in the sea, not the time since Earth's formation). It is estimated that the entire volume of the world's oceans flushes through rocks of the midocean ridges in about 8 million years.[4]

Not all the infiltrating water escapes, however. Having entered into labyrinthine passages and forged chemical bonds with minerals in the basalt, some is now locked into the ocean crust for the long term. As it happens, this accidental entrapment of water is one of the most essential components of Earth's tectonic system. A subducting slab carries the stowaway water from its youth as it descends into the mantle. The cold slab slowly warms, and when it reaches depths of about 30 miles, it finally sweats out this ancient seawater. We tend to think of the water cycle as a relatively short-term phenomenon; the average molecule of water stays in the atmosphere for about nine days; the residence time of water even in the largest lakes, like Superior, is a century or two; deep groundwater may be stored for a millennium. But there is a 100 million-year water cycle that involves the interior of the Earth, and adding water to the mantle is in fact the critical step in the recipe for continental crust.

In the presence of water, the otherwise solid rock in the wedge of mantle above a subducting slab will melt at significantly lower temperatures than it normally would, in the same way that salt lowers the melting temperature of ice on a sidewalk. This "water-assisted" melting is both creative and destructive: it ultimately forges new continental crust but does so via some of the deadliest volcanoes on Earth, which form on the overriding plate in the subduction zone, directly above the spot where the downgoing slab gives up its long-sequestered water. The volcanoes typically form an arcuate chain—a broad C shape that reflects the curvature of a subduction trench on a spherical Earth, like the crescentic shape of a dent in a ping-pong ball. Where the upper plate is also basaltic ocean crust, the volcanic chain is called an *island arc*. Examples include Japan, Indonesia, the Philippines, the Aleutians, and the north half of New Zealand. If a subducting plate dives beneath a continent, the resulting volcanoes form a *continental arc*, like the Cascades and Andes (see figure 8).

In both arc settings, the water-generated mantle melt must make its way through the upper plate en route to the surface. The magma may be stalled by the rigidity of this lid of obstructing crust and, while ponded there, partially melt it. Just as at midocean ridges, the low-melting temperature components will be most readily extracted, giving rise to new magmas that are even richer in silica and less like the mantle than basalt is. Through many such cycles of such smelting, crust that is progressively more "evolved" is generated, ultimately yielding granite, the lightweight raw material for the continents. Plate tectonics on the modern Earth is an extraordinary system. The creation, maturation, and eventual destruction of ocean crust are all necessary for the genesis of continental crust—a perfect samsara cycle of birth, death, and reincarnation.

## MOUNTAIN TIME

An oceanic subduction zone will function smoothly (though not necessarily aseismically) as long as the crust entering the trench is thin and dense enough to slide into the mantle. But if the slab pulls in "undigestible" things like ocean crust that is too hot or too thick, or a lumpy old island arc—or an unsinkable continent—traffic comes to a halt. And if the upper plate is a continent, a major pile-up is unavoidable, and a mountain belt begins to grow. The loftiest mountains on Earth, like the Himalaya now, and the Alps, Appalachians, and Caledonides in their day, form when a long-lived subduction zone has swallowed an entire ocean basin, and two continents are set on a collision course.

How long does it take to raise a mountain belt? In the case of the Himalaya, the seafloor spreading history recorded by marine magnetic anomalies makes it possible to track India's headlong rush from its place in the ancient southern continent of Gondwandaland in the late Cretaceous to its current position as part of Asia.[5] Pulled northward by subducting ocean crust, India traversed about 2500 km (1500 mi) in 30 million years (an impressive average pace for such a marathon, at more than

8 cm (3 in.) a year), before first running into Asia about 55 million years ago. Since then, the Himalayan range has risen as the northern part of India has wedged itself beneath Asia, and the crust of both continents has also thickened vertically through faulting and folding. As the convergence continues, a wave of deformation is propagating outward from the original point of contact, both north and south, progressively widening the welt of uplifted and contorted crust.

Before the emergence of plate tectonic theory in the 1960s, mountain belts were difficult to explain. Many geologists recognized that the buckled, crumpled strata typical of montane regions required horizontal compression, but the motive force behind this was difficult to understand under the prevailing assumption that continents were fixed in place. A nineteenth-century Austrian geologist, Eduard Suess (grandfather of Hans, who would document the dilution of atmospheric $^{14}C$ by "dead" carbon from fossil fuel burning), recognized that many of the rocks in the Alps had formed on the seafloor and had somehow been elevated to their present positions. He postulated that Earth's mountains were akin to the wrinkles on a raisin, ridges formed by shrinkage as the result of steady cooling and contraction of the planet—a notion consistent with Lord Kelvin's views of the thermal evolution of the interior.

The art critic, intellectual polymath, and alpine enthusiast John Ruskin, a contemporary of Suess's, also had an intuitive sense that mountains are not static, eternal monuments but records of dynamic events. For Ruskin, however, the morphology of the Alps evoked liquid fluidity rather than desiccated fruit: "There is an appearance of action and united movement in these crested masses, nearly resembling that of sea waves . . . fantastic yet harmonious curves, governed by some grand under-sweep like that of a tide running through the whole body of the mountain chain."[6] He also recognized that that these "harmonious" shapes represented the countervailing effects of the "elevatory force *in* the mountain" and "sculptural force of water *upon* the mountain." But how efficiently do these opposing forces act?

The highest Himalayan peaks, at elevations of 9000 m (29,000 ft), lie where the coast once was. So it might seem logical to estimate their growth rate simply by dividing their height by 55 million years, which yields a positively underwhelming uplift rate of 0.015 cm (0.006 in.) per year. But this calculation is a gross underestimate of the actual rate of mountain building—because as soon as tectonic forces start constructing mountains, the highly efficient erosion crew arrives to begin demolition. So we need to find ways to measure these opposing processes independently.

Today, surface uplift can be measured in nearly real time thanks to high-precision global positioning system (GPS) satellites. In the highest part of the Himalaya, the Tibetan Plateau, GPS-based uplift rates averaged over a decade are in the range of 2 mm (0.1 in.) per year. This is about an order of magnitude slower than the tectonic convergence rate (around 2 cm, or 1 in., per year)[7] and reflects a fairly typical ratio of vertical to horizontal deformation in the crust. But the instrumentally measured uplift is more than 100 times faster than a long-term estimate that ignores the effects of erosion. How can we know whether modern satellite-based estimates are representative of uplift rates over longer periods of geologic time? As the "roof of the world" is rising, its top stories are constantly being removed, in a process geologists call *exhumation*. What were once the subterranean levels now have high penthouse views. To reconstruct long-term uplift rates, we need to know how many floors have been dismantled, and how quickly.

There are several ways to calculate how much additional rock once existed in the airspace above a mountain belt. One is to ask the rocks now at the surface how deep they used to be at a certain time in the past. This can be done using a technique called *fission track dating*, developed largely by oil companies to reconstruct the thermal histories of sedimentary rocks to predict whether they are likely to produce petroleum or natural gas (sediments need to have been warm enough that their organic matter got properly "cooked," but never so hot that it all burned off).

Fission track dating makes use of the fact that the more abundant isotope of uranium, $^{238}U$, is not just radioactive but also has an unstable nucleus that splits itself in spontaneous fission events at a known rate. Under high magnification, uranium-bearing minerals including zircon (the darling of geochronology) and apatite (the mineral in teeth and bones) retain a visible record of these high-energy events in the form of damage zones or "fission tracks." Each uranium-bearing mineral has a particular temperature above which the crystal lattice can heal itself and erase these scars, like an Etch-a-Sketch that has been well shaken. Below this temperature, however, the tracks will remain etched in the crystal. So, by counting the density of fission tracks in a given volume of a mineral, it is possible to determine how long it has been since it cooled through a certain temperature (and depth) in the crust. Fission track *thermochronology* for Himalayan rocks shows that modern uplift rates based on a few decades of satellite data are in fact consistent with uplift over geologic timescales.[8]

## REMAINS OF THE DAY

Another approach to estimating how much has been trimmed from mountains by erosion is to look at the volumes of sediment that have accumulated at their feet, like snippets of hair on a barbershop floor. In the Himalaya, most of the erosional detritus has accumulated in two gigantic sandpiles on the seafloor: the Indus and Bengal submarine "fans," where the Indus, Ganges, and Brahmaputra Rivers have dumped sediment for the past 50 million years. On Marie Tharp's maps, the Indus and Bengal fans are long tongues lolling far out onto the floor of the Indian Ocean. The Bengal fan is the largest in the world. From the mouth of the combined Ganges-Brahmaputra on the coast of Bangladesh—which itself is made entirely of sediment shed from the mountains—the fan extends 3000 km (1800 mi) southward. If superimposed on the continental United States, the Bengal fan would stretch from the Canadian border to Mexico, and for almost half that distance, it is more than 6.5 km (4 mi) thick.

Drilling and geophysical exploration of the Indus[9] and Bengal[10] fans have revealed an upside-down, impressionistic record of the unroofing of the Himalaya, with the disaggregated remains of rocks that were at the top of the mountains in their infancy now forming the lowest layers in the immense mass of deep-sea sediment. The total volume of the Bengal fan alone, an estimated 12.5 million $km^3$ (3 million $mi^3$),[11] is greater than the present-day volume of the crust of the Tibetan Plateau above sea level.[12] That is, more rock has been removed from the Himalaya by erosion than forms the towering modern range. This fact makes the seemingly simple question posed by Hutton (and Dylan)—How long does it take to wear mountains down?—more difficult to answer. Which mountains are we talking about? The Himalaya have existed for 55 million years, but today's mountains are not the same as the mountains whose ruins lie on the floor of the Indian Ocean.

The ephemeral nature of mountains—or any landscape—is one reason that unconformities in the rock record, like Hutton's famous outcrop at Siccar Point, hold such fascination. Unconformities preserve buried topography and thus provide glimpses of the long-vanished terrain of earlier eons. Wisconsin's Baraboo Hills region, a mecca for geology field trips (and home of the late Ringling Bros. and Barnum & Bailey Circus, the "Greatest Show on Earth"), represents one of the most remarkable examples of paleotopographic preservation anywhere in the world. This Precambrian mountain belt, formed about 1.6 billion years ago, was buried by hundreds of feet of marine sediments when early Paleozoic seas washed over what is now the Great Lakes region. Today, erosion of these Paleozoic rocks has reached a stage in which the unconformity between the Precambrian and Paleozoic worlds is exposed in many places. The long-hidden mountains are being unburied, or exhumed, and the modern land surface closely approximates that of the late Proterozoic. Interestingly, this ancient landscape was the inspiration for two great environmental thinkers: John Muir, whose family immigrated to the area from Scotland when he was a young boy, and Aldo Leopold, whose *Sand County Almanac* is set in the shadow of the primordial Baraboo Hills. There are much older rocks, and deeply eroded roots of older mountain belts in other places (even Wisconsin), but the Baraboo Range represents some of the oldest preserved *topography* on Earth—a Great Show indeed.

## THE HILLS ARE ALIVE

The sediments shed from the Himalaya tell us that while there has been some variation in uplift and exhumation rates over time, on average these rates fall within the range of estimates from both GPS observations and thermochronologic approaches like fission track dating. This is a comfortingly uniformitarian result; Lyell would be pleased. The great heaps of sediment also underscore an amazing fact about the Earth: that the speeds of tectonic processes, driven by the internal radioactive heat of the Earth are, by happy coincidence, about evenly matched[13] by the tempo of external agents of erosion—wind, rain, rivers, glaciers—powered by gravity and solar energy. In the barbershop analogy, it is as if the hair on a customer's head keeps growing as fast as the barber can cut it. And while the tectonic growth and erosional trimming of mountains both proceed at an average pace that is deliberate, they are not so slow as to be beyond our perception.

That additive and subtractive topography-forming processes are so commensurate is one of Earth's extraordinary attributes. The landscapes of other rocky planets and moons look alien precisely because these worlds lack such a balance in the rates of creative and destructive topographic agents. On Earth, if tectonics far outpaced erosion, mountain plateaus would persist longer, creating vast areas of alpine habitat. If erosion outstripped tectonics, the continents would be lower but more rugged, and rivers would carry greater volumes of sediment to continental shelves, dramatically changing the nature of coastal regions. In either case, life on land and in the sea would face different natural selection pressures, and evolution would likely have followed alternative routes. However, life itself can alter the processes that shape topography: there is strong evidence that colonization of land by plants in early Silurian time (ca. 400 million years ago) slowed global erosion rates and led to the emergence of rivers with well-defined channels.[14] (It has taken humans only a few centuries to reverse that trend; by some estimates, modern erosion rates—accelerated by deforestation, agriculture, desertification, and urbanization—are orders of magnitude higher than geologic averages.[15])

Remarkably, the pace of biological evolution is well matched with the rates of tectonic and surface processes over geologic timescales. This is particularly evident in the Hawaiian Islands, which have formed in sequence from northwest to southeast, as the Pacific Plate has passed over a deep-seated "hotspot" where mantle rock wells up and melts through decompression. A study of biodiversity on each island over time shows that adaptive radiations—bursts of evolutionary innovation—coincided with the growth of each island through volcanism, then leveled off as erosion gained the upper hand, reducing an island's area and elevation range.[16] And of course, Darwin's original insights about evolution came from the diversity of species on the equally youthful Galapagos Islands (whose age was, however, not known to him). One could imagine an alter ego planet where surface morphology changed too quickly for evolutionary adaptation of macroscopic life, like a ballet orchestra that is playing so fast the dancers can't keep up. Fortunately, all members of the Earth ensemble—volcanoes, raindrops, ferns, and finches—perform in synchrony.

## RAIN AND TERRAIN

A closer look at how mountains develop reveals even subtler relationships between tectonics and erosion—and further complicates the Hutton-Dylan riddle. First, rates of erosion depend on weather and climate, and tectonic topography can change both. Like air travelers allowed to take only small amounts of liquid past security checkpoints, air masses are forced by mountains to drop their moisture as they pass over the crest line, which creates rain shadows on leeward slopes and leads to asymmetrical rates of erosion across the mountain belt. In India, the intensity of the annual monsoon is directly linked with the existence of the Himalaya, leading to ferocious erosion in the steep foothills. The Tibetan Plateau, meanwhile, owes its height in part to the arid conditions the mountains themselves have created. However, aridity leads to lack of vegetation, which makes slopes more vulnerable to gravitational failure in landslides. As they grow, then, mountains create their own complex climate systems, which in turn shape their future evolution.[17]

Great mountain belts like the Himalaya can even change global climate. During the Cretaceous Period, before the collision of India and Asia, Earth had a hothouse climate, with no glaciers or ice caps. An inland sea covered the Great Plains region of North America and lapped up against western Minnesota. Seafloor spreading had been unusually fast for about 40 million years, leading to higher-than-average amounts of volcanic carbon dioxide ($CO_2$) in the atmosphere. Some dinosaurs even lived at high arctic latitudes. Starting in the early Cenozoic Era, at about the same time that the Himalaya began to rise, Earth's climate began the long-term cooling trend that has characterized the last 50 million years. Many geologists believe there is a causal connection between this cooling and the creation of high topography in the Himalaya. In particular, the chemical weathering of rocks by rainwater is, over geologic timescales, an important mechanism for drawing carbon dioxide ($CO_2$), the most abundant greenhouse gas, out of Earth's atmosphere (see figure 9 and chapters 4 and 5).

In the absence of human activity, $CO_2$ comes mainly from volcanic exhalations. When $CO_2$ mixes with water vapor in the atmosphere, it forms a weak acid (carbonic acid, $H_2CO_3$) that is effective at dissolving rocks over time. Many crustal rocks contain calcium, which is then carried in solution by rivers to the world's oceans. In the sea, organisms ranging from corals and

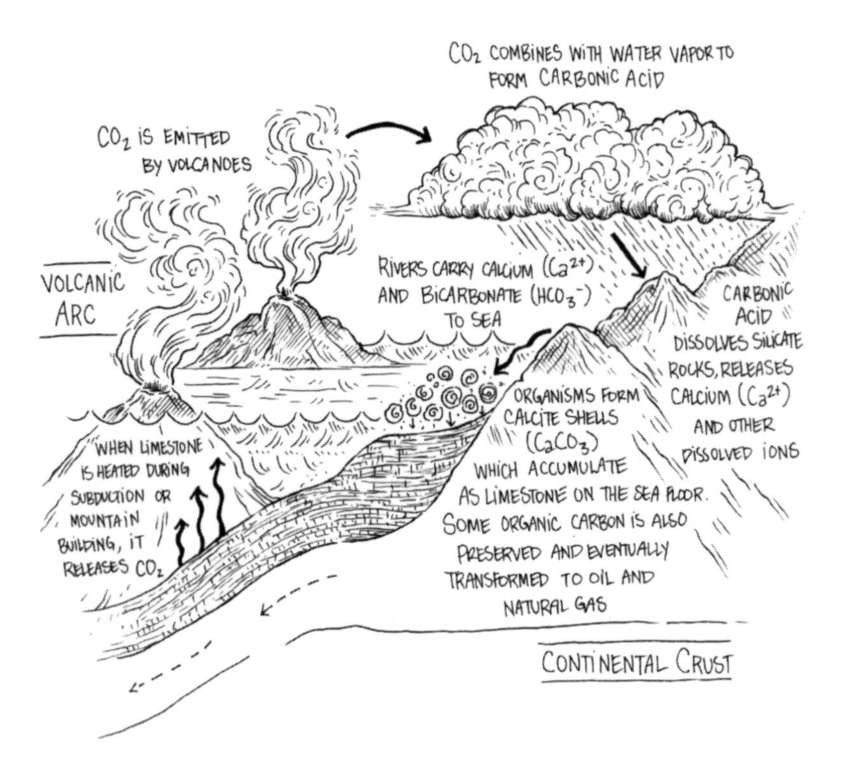


FIGURE 9. The long-term carbon cycle; the weathering of mountains regulates atmospheric $CO_2$

starfish to single-celled zooplankton use this calcium, together with bicarbonate ($HCO_3^-$) to form shells and exoskeletons made of calcite ($CaCO_3$). This whole process can be written in simplified form as a sequence of chemical reactions:

Rock weathering → Ions dissolved in rivers → Formation of limestone

$$CO_2 + H_2O + CaSiO_3 \rightarrow Ca^{2+} + 2HCO_3^- + SiO_2 \rightarrow CaCO_3 + SiO_2 + CO_2 + H_2O$$

| Combine to make acid | Simplified composition of igneous rock | Calcium in solution | Bicarbonate | Calcite secreted by marine organisms | Silica (used by other organisms such as sponges) |
|---|---|---|---|---|---|

But the most critical step, from the perspective of long-term climate modulation, is that when calcite-secreting organisms die, their mineral remains rain down to the seafloor to form limestone, locking up atmospheric carbon dioxide in solid form, where it remains for tens of millions of years.

This is Earth's long-term carbon sequestration program—a greatly underappreciated ecosystem service—and it is more efficient at times when lots of fresh rock surfaces are being made available for chemical weathering, such as during the construction of a Himalayan-scale mountain belt. The growth of the Himalaya, then, influenced not only local and regional weather patterns, but climate and even topography at a global scale, ultimately helping push Earth into the Ice Age, when glaciers and ice caps reshaped landscapes all over the world.

## PEAK PERFORMANCES

Another even subtler, and counterintuitive, link between erosion and mountain building involves the way that a mountain belt interacts with Earth's mantle. As mountains form owing to tectonic collision and crustal thickening, the added weight of so much rock piled up in one place causes the weak (though solid) upper mantle—called the *asthenosphere*—to be displaced, like the water beneath a heavily laden ship. But once a mountain belt stops growing (as in the case of the young but no longer tectonically active Alps), erosion gets the upper hand and reduces the weight of the crust. This causes the displaced mantle to flow back into place and the mountains to rise in elevation, like a ship emptied of cargo (such *isostatic rebound* also occurs in areas previously covered by thick sheets of glacial ice[18]). In this way, erosion paradoxically helps raise mountains.[19]

Throughout the life of a mountain belt, then, crustal deformation, climate, erosion, and mantle displacement perform a languid interactive dance in which each player influences the motions of the others. But on occasion, their slow-motion choreography is disrupted by sudden jumps and jetés. Charles Darwin, who experienced a great earthquake in Chile while on the *Beagle* expedition, was perhaps the first to speculate that these destructive events may in fact help build mountains over time, even though the cause of earthquakes—sudden slip on faults—was not fully understood at the time. Noting a bed of "putrid mussel-shells" that had been heaved 3 m (10 ft) above the high-water mark by the earthquake, Darwin speculated that older seashells he found at elevations up to 180 m (600 ft) had been brought there by "successive small uprisings, such as that which accompanied or caused the earthquake of this year."[20] As usual, Darwin was right.

Unlike most geologic processes, which are difficult to study because they elapse slowly, earthquakes can be experienced in real time, but they occur at inaccessible depths. No one has ever directly witnessed what happens on a fault surface deep in the crust when an earthquake occurs, but a century of seismologic research integrating elastic wave theory, experimental rock mechanics, and analysis of modern and ancient fault zones makes it possible to extract many types of quantitative inferences from the squiggly lines on a seismogram. The largest earthquakes are magnitude 9 (M9) *megathrust* events in subduction zones, like those that took place in Indonesia in 2004 and Japan in 2011. Such events can accomplish in minutes what would take hundreds of years at background tectonic rates.

In the devastating tsunami-inducing 2004 Sumatra earthquake, an astonishing 1100 km (700 mi) of the plate boundary was activated.[21] The underwater rupture propagated northward from its origin over a period of 10 hellish minutes at a velocity of more than 1.6 km/s (1 mi/sec),or 6900 km/h (4300 mi/hr). All along this distance, the Sunda plate, carrying Indonesia, lurched an average of 20 m (65 ft) westward, a displacement equivalent to about 1000 years' worth of normal plate motion. As each successive segment of the plate boundary slipped, powerful seismic waves—the cause of the actual ground-shaking in an earthquake—were generated, moving outward in concentric circles like ripples on a pond, at speeds of 3 to 5 km/s (2 to 3 mi/sec). Clocking these rates is of more than academic interest; while rupture fronts and seismic waves are fast, electromagnetic waves that transmit digital information are still faster. In Indonesia, Japan, and other areas of high seismic risk, cellphone earthquake and tsunami alert systems have been implemented in the hope that a few critical seconds of warning may help save lives in future events.

While we can't predict exactly when or where great earthquakes will occur, we can say with utter certainty there will be many more. Global instrumental seismic records now span almost a century and show that, on average, an M9 megathrust earthquake can be expected along one of Earth's subduction zones every few decades. Worldwide, on all types of faults, there are typically one or two M8 and tens of M7 events each year.[22] Building earthquake-resistant housing in seismically active regions should be one of the world's top humanitarian priorities. In the twenty-first century, an M7 earthquake should not cause 100,000 deaths, as the January 2010 Haiti quake did. Our surprise and shock when yet another earthquake devastates a city and claims thousands of lives is almost medieval.

## FAULTY LOGIC

For decades, geoscientists thought that faults accommodated deformation of the Earth's crust in two distinct modes with radically different tempos: fast and furious (meters per second) during earthquakes, but slow and steady (centimeters per year) the rest of the time. Furthermore, there seemed little common ground between the physical phenomena that occurred on fault zones at such different timescales. As a result, seismologists who study earthquakes, and geologists who study the gradual tectonic processes that build mountain belts ("structural" geologists, like me), have traditionally been two distinct academic clans. More recently, however, the two fields have begun to converge. In the late 1980s, a distinctive glassy rock type with the cumbersome name *pseuodotachylyte*, sometimes found in ancient fault zones, was shown to be the product of localized frictional melting, which could have happened only at slip rates of meters per second—that is, during earthquakes. This discovery has made it possible to observe directly the physical consequences of seismic slip on rocks that were at the focus of an earthquake. And since the start of this millennium, a new generation of seismic arrays, combined with high-resolution GPS monitoring of ground motion and more powerful data processing, has led to the discovery that faults have a much wider spectrum of behaviors than previously thought.

Between long-term "creep" that occurs at background tectonic rates, and conventional earthquakes that occur in seconds to minutes, geoscientists have now documented intermediate events called *slow earthquakes* that elapse over days to weeks, generating very low frequency tremors that had previously been dismissed as noise. In contrast with earthquake rupture speeds of kilometers per second, these events propagate along a fault zone at a sedate—even walkable—rate of 16 to 32 km (10–20 mi) per day. Oddly, some of them then double back on themselves, reversing their sense of propagation at a slightly higher velocity than they took on their outward journey,[23] like a hiker who quickly retraces her footsteps to pick up a dropped mitten. Stranger still, slow-slip events on some faults zones recur at regular, but cryptic, intervals. On the Cascadia subduction zone off the coast of Washington State and British Columbia, for example, slow earthquakes follow a 14-month cycle whose significance is not understood.[24]

The causes and consequences of slow-motion seismicity are not yet clear. Many geologists think these episodes could be related to fluids percolating through deforming rocks, and if so, mineralized fractures in ancient rocks, called *veins*—the source of many metallic ores—may in fact be records of ancient slow earthquakes. While this concept is intriguing, a more important question is the relationship between sluggish earthquakes and sudden, devastating ones. Do slow earthquakes help reduce stress on faults by relieving it in increments, or do they presage larger, potentially catastrophic events?[25] Studies of fault zones around the world—in the western United States, New Zealand, Japan, Central America—suggest that the answer may not be the same for all depths and fault zones, which is an unsettling conclusion. It also seems likely that faults have secret habits on timescales of centuries to millennia that as yet fall outside the range of our observational abilities.

## GOING DOWNHILL

Just as the building of mountains is generally unhurried but sometimes impulsive, their demolition alternates between continuous and quantized. We humans think we are glimpsing eternity in rocky alpine landscapes, but in fact they inspire thoughts of infinity in us just at the point they themselves sense mortality. Majestic peaks and magnificent palisades are simply what remains, for now—the provisional results of the latest cuts by a team of obsessive sculptors: water, ice, and wind in artistic collaboration with gravity. Yet we are shocked when a rockfall scars a cherished cliff face in Yosemite or disfigures the iconic Old Man of the Mountain in New Hampshire. Some research in the field of *geomorphology* (the study of landscape evolution) suggest that in mountainous areas, episodic landslides and other types of large-scale slope failure are the single most important mechanism of erosion, while rivers (previously thought to be the prime movers) just tidy up after them in the intervening decades to centuries.[26]

Earthquakes, of course, can trigger landslides, and while they generally help construct mountains, the landslides they unleash may, in certain cases—as in the tragic 2008 Wenchan earthquake in China—actually negate the tectonic uplift they cause.[27] In other words, the creation and destruction of montane landscapes are intimately linked, and both may be dominated less by long periods of uniformitarian boredom than short periods of real-time terror.

There is geologic evidence for ancient slope failures that are far larger in magnitude than any experienced in human history—so extreme that they seem like implausible scenes in a bad apocalyptic sci-fi film. For example, about 73,000 years ago, the catastrophic collapse of the flank of a volcanic island in the Cape Verde archipelago off the west coast of Africa generated a tsunami that hurled 90 t (1 t = 1000 kg = 2200 lb) boulders 180 m (600 ft) up the side of another island 50 km (30 mi) away.[28] And while most people are aware that Yellowstone lies above a sleeping supervolcano that has exploded in unimaginably gigantic eruptions, a mountain just outside the park records an ancient catastrophe that is even more terrifying. Heart Mountain, Wyoming (the site of an Japanese-American internment camp in World War II), is part of a 1.6-km (1-mi)-thick rock slab the size of Rhode Island that slid more than 50 km (30 mi) across a surprisingly gentle slope in 30 minutes—that is, at highway speeds—perhaps aided by superheated gases at it base.[29] These outsized events remind us that our short window of observation has not exposed us to the full range of Earth's behavior and suggest that what we consider "normal" landscape processes may actually be more like the activity of a relief crew attempting to restore infrastructure after a disaster. Charles Lyell would not like this idea.

## UNCHARTED TERRITORY

Understanding the lingering effects of sudden topographic change is important because we ourselves are now agents of geomorphic catastrophe. The coal-mining practice of "mountaintop removal"—a deceptively surgical term—moves volumes of rock that rival the largest natural disasters. In parts of Appalachia, old topographic maps have simply become irrelevant. A 2016 study of the mutant landscape of southern West Virginia determined that since the 1970s, some 6.4 $km^3$ (1.5 $mi^3$) of "overburden" waste rock has been moved from mountain summits and dumped in the upper reaches of stream valleys.[30] That volume is on par with the amount of sediment that the Ganges and Brahmaputra—two great rivers draining the mightiest mountains on Earth—carry to the Bengal fan in a decade. And this is in just southern West Virginia.

The effects of such massive derangement of the landscape will be wide ranging and long lasting. Where trees once anchored soil on top of bedrock, piles of broken mine waste, hundreds of feet thick, now mantle the slopes. In nature, rivers shape hill slopes until they reach a stage of being *graded*—just steep enough that their flow velocities can keep pace with the sediment supplied by the valley. In the devastated valleys of Appalachia, the small infilled upland streams will seek valiantly to process the colossal volumes of waste rock. Estimating how long this will take is difficult because there is almost no geologic analog for such a profound state of disequilibrium, but hundreds of thousands of years is probably a conservative estimate. Predictions about the short- and long-term effects on surface and groundwater chemistry and the fate of native plants and animals are equally sobering. And the psychological effects on humans left in the shadow of the decapitated mountains is beyond quantification.

Worldwide, humans now move more rock and sediment, both intentionally through activities like mining, and unintentionally by accelerating erosion through agriculture and urbanization, than all of Earth's rivers combined.[31] It can no longer be assumed that geographic features reflect the work of geologic processes. In a matter of years, the Chinese government has radically altered the map of the Spratly Archipelago in the South China Sea by scraping coral reef material from the seafloor to create new islands, in a dystopian counterpoint to the formation of Surtsey. In southern England, rates of retreat of the famous chalk cliffs have accelerated from inches per year to feet per year as a result of human changes to the shoreline combined with encroaching seas and increased storminess due to climate change.[32] The Nile Delta is sinking 2.5 to 5 cm (1 to 2 in.) per year as a result of being starved of sediment by the Aswan and other dams.[33] Coastal Louisiana is losing an acre of land *per hour* as a result of a "perfect storm" of unintended consequences: continent-scale engineering of the Mississippi channel has dramatically reduced sediment supply at the same time that oil and gas withdrawal has caused the land to subside—all while the sea inexorably rises (an indirect result of... oil and gas consumption).[34] Meanwhile in Oklahoma, we have reawakened long-dormant faults and induced earthquakes through deep-subsurface injection of wastewater generated by the practice of hydrofracturing for oil and gas extraction.[35]

The unprecedented scale of human changes to the planet's topography is one of the arguments for the concept of the Anthropocene, a new division of the geologic timescale marked by the emergence of humans as a global geologic force. We are literally changing the configuration of the continents and remaking the world map. But does this matter on a planet that has seen so many geographies, constantly erasing old worlds and replacing them with new ones? It doesn't to the Earth itself, which will eventually remodel everything according to its own preferences, either gradually or catastrophically. Over human timescales, however, our disruption of geography will haunt us. Soil lost to erosion, coastal areas claimed by the sea, and mountaintops sacrificed on the altar of capitalism won't be restored in our lifetime. And these alterations will set in motion a cascade of side effects—hydrologic, biological, social, economic, and political—that will define the human agenda for centuries. In other words, thoughtless disregard for the work of the geologic past means we cede control of our own future.

In 1788, when James Hutton saw the unconformity at wave-swept Siccar Point, he imagined the eons it would take to remove a mountain and concluded that geologic time was infinite. More than 200 years later, we can clock the growth and destruction of mountains. The famous unconformity, which separates Silurian rocks from Devonian ones, represents not eternity but about 50 million years, which is plenty of time to build and demolish a mountain belt—for continents to collide, faults to creep and sometimes lurch, raindrops to sculpt, peaks to crumble, mantle rock to flow. Today we can even observe the workings of the solid Earth in real time. We find that the planet's natural pace is not so far outside our own experience, and that in fact this old orb has a wide repertoire of tempos, including some that are breathtakingly swift. Studying the habits of the solid Earth teaches us to respect the power of both incremental change and episodic catastrophe to transform the face of the globe.

The lingering nineteenth-century belief that Earth changes only slowly has lulled us into thinking that it is impassive and eternal, that nothing we do could alter it significantly. That notion has also caused us to view Earth's intermittent adjustments—the creation of a new volcanic island, a magnitude 9 earthquake—as aberrations, when in fact these events are business as usual for the planet. We are big enough now to scratch and dent the Earth, scar, and abrade it, but we ourselves will have to live with the damage. Earth, meanwhile, will continue to make slow repairs, punctuated by sudden renovation projects that will clear away our proudest constructions.

CHAPTER 4

# CHANGES IN THE AIR

Here feel we not the penalty of Adam,
The seasons' difference, as the icy fang
And churlish chiding of the winter's wind,
Which, when it bites and blows upon my body,
Even till I shrink with cold, I smile and say
"This is no flattery. These are counselors
That feelingly persuade me what I am."
. . . . . . . . . . . . . . . . . . . . . . . . . . . . . . . . . .
And this our life, exempt from public haunt,
Finds tongues in trees, books in the running brooks,
Sermons in stones and good in every thing.
I would not change it.

—WILLIAM SHAKESPEARE, 1599.
*AS YOU LIKE IT*, ACT 2, SCENE 1

## COLD COMFORT

Many of the geographic features in Svalbard had no formal names until the late nineteenth century, and in the area where I did my graduate-school fieldwork, some of them were christened in honor of geologists of the day. A lofty peak was named for Jöns Jacob Berzelius, the "Father of Swedish Chemistry" and a pioneering mineralogist. A relatively sheltered valley with a half-dozen picturesque glaciers was dubbed Chamberlindalen, for T. C. Chamberlin, a Wisconsin geologist who first mapped glacial deposits in the upper Great Lakes region. A windy point jutting into the Arctic Ocean is called Kapp (Cape) Lyell for the great evangelist of uniformitarianism.

My own work in Svalbard in the 1980s was itself something of a throwback to the nineteenth century: creating a geologic map of the region by delineating unnamed rock units and charting their extent, collecting samples for analysis, and making a provisional interpretation of the area's geologic history. This kind of reconnaissance had been finished decades before in most of the rest of the world.

The base maps on which we would plot our geologic observations were enlargements of beautiful hand-drawn charts from the 1920s and '30s. I loved their graceful, slanting fonts and the way the lettering curved to conform to the arcs of glaciers and coastlines. But the *contour interval* (the spacing between lines of constant elevation) was a gap-toothed 50 m (about 170 ft)—a very coarse sieve through which a considerable amount of topography could fall. So in the field we would make notes on aerial photographs that the Norwegian Polar Institute had taken in the 1930s and '50s (interrupted by the desperate war years, when Norway was fighting for its existence, and even remote Svalbard had U-boats lurking in the fjords). We'd then transfer the information to the maps each evening, by the light of the midnight sun. Air photos like these—now largely superseded by satellite imagery—came in overlapping pairs, which when viewed with stereoscopic glasses would make topographic features pop out in exaggerated 3-D, like tableaus seen through an old "Viewmaster" toy. (Some seasoned field geologists could achieve the same effect by relaxing and slightly crossing their eyes, though I never did acquire this skill). We quickly learned that we needed to be careful when plotting our locations on the air photos, because the positions of glacier margins were commonly farther up-valley than they were on the old images. These were the first hints that time was coming to "timeless" Svalbard.

In subsequent years, I was lucky to do geologic work in stunning glacial landscapes elsewhere in Svalbard, as well as the Canadian arctic, but I didn't revisit Kapp Lyell until 2007, exactly 20 years since I had last seen it. Returning to a place that I had studied with such intensity at an earlier time in my life threw into stereoscopic relief how much I had changed in that time, having experienced marriage, an academic career, the birth of three sons, the death of a spouse. However, I expected the landscape whose contours I remembered so well to be more or less the same. It was eerie to find our old campsite, with boulders we had used to anchor the cook tent, exactly where we had left them. But almost everything else was dramatically altered. Our group had been able to arrive by boat before the middle of June—weeks earlier than was possible in the 1980s—because the sea ice had not even reached the southern part of Svalbard that year. (In fact it was the first time in history that the fabled Northwest Passage was also ice-free). This meant that polar bears, which used to spend summers idly drifting with the ice floes and dining on seals, and had never given us much trouble before, were walking around on land, hungrily eyeing up geologists. Even more disturbingly, all the familiar Chamberlindalen glaciers, once white and plump, had become sickly gray ghosts of themselves, shrunken far up into their mountain headwalls . For almost two decades, I had been presenting the evidence for climate change in my university classes and had facts and talking points I could recite in my sleep. But seeing the shocking alteration of a place that I knew so intimately was like arriving at what one expects will be a joyful reunion of old friends—and finding them all deathly ill. The name Kapp Lyell now seemed a mocking irony; this was not uniformitarianism.

Time, which had for so long left Svalbard in its Ice Age slumber, was returning with a vengeance.

## AIR OF MYSTERY

The changes in Svalbard's glaciers make clear that even a remote place near the top of the globe is connected to the rest of world through the atmosphere. The concentric layers of the Earth scale remarkably well to the parts of a peach: the iron core corresponds to the pit, the rocky mantle to the flesh of the fruit, the crust to the skin. The atmosphere, in turn, is proportionally as thick as the exterior fuzz, extending 480 km (300 mi) above the surface, though most of its mass is concentrated in the lowest 16 km (10 mi). Ubiquitous but mostly invisible, the atmosphere is one of the great amenities provided by this accommodating planet. In contrast to the carbon dioxide-dominated atmospheres of Venus and Mars, which are little more than stagnant volcanic exhalations (crushingly heavy on Venus, mostly lost to space on Mars), Earth's mix of nitrogen and oxygen with just trace amounts of $CO_2$ is anomalous and marvelous. Understanding its deep history can help put modern rates of atmospheric and climate change into some kind of perspective. The story of the atmosphere is bound up inextricably with the story of life; life itself crafted the modern atmosphere—in a sense, wrote its own chemical constitution. Life has governed stably for much of the geologic past, but occasionally, even a sophisticated system of biogeochemical checks and balances has not been enough to prevent atmospheric revolution and ecological catastrophe.

How do we know anything about ancient air? For the past 700,000 years, we have a direct record of its composition from gas bubbles trapped in ancient snow and then preserved as polar ice (more about that in the next chapter). But where can we look for information about something so evanescent over longer timescales? Counterintuitively, rocks—the antithesis of all that is airy—have much to tell us about the atmosphere. In particular, they reveal that the modern atmosphere is at least the fourth major version of Earth's rarefied outermost layer. Contrary to Hutton's and Lyell's views of an Earth in a state of perpetual, but directionless, cycling, the history of the atmosphere is a *Bildungsroman* about a planet reinventing itself as it matured. Like the air in a building—smoky, moldy, well ventilated, or heavy with cooking smells—Earth's atmosphere reveals much about the habits of its residents. For at least 2.5 billion years, the biosphere has altered the atmosphere at the planetary scale, and conversely, every mass extinction and major disruption in the biosphere has coincided with dramatic changes in the composition of the atmosphere. While the evolution of the air is linked with that of the solid Earth through volcanism, rock weathering, and deposition of sediments, the atmosphere is generally much nimbler than the tectonic system, capable of quicksilver transformation. A deep dive into the history of Earth's invisible envelope may give us a new appreciation for each breath we take.

## FIRST BREATHS AND SECOND WIND

Earth's very first atmosphere was probably rocky—that is, heavy with pulverized and vaporized rock from a constant barrage of high-velocity extraterrestrial objects. Apart from the celebrated Jack Hills zircons (chapter 2) there is no known Earthly record of the planet's first 500 million years. The only extensive source of information for this interval—the *Hadean* Eon, the "hidden" or "hellish" time—are samples gathered by astronauts and cosmonauts from the Moon. Its familiar, scarred face, with rocks as old as 4.45 billion years blanketed by shattered rock fragments (the lunar *regolith*), attests to a violent regime of relentless impacts as debris left over from the formation of the solar system pummeled the young inner planets.

This debris likely included not only rocky and metallic meteorites but also icy comets carrying water from orbits beyond Neptune to the infant Earth, which would have had only limited native supplies of its own, given its proximity to the Sun. In any case, the Jack Hills zircons suggest that within 100 million years of its formation, some water already existed on Earth's surface or at least in the shallow crust—the earliest hints of what would become its signature attribute. Yet we know from the Moon's surface that heavy bombardment continued until at least 3.8 billion years ago, when the great, dark *maria* basins of Galileo—themselves giant craters—formed. In Hadean time, the Moon was even closer to the Earth that it is today, and there is every reason to think that Earth must have been similarly pelted for its first 700 million years. It is probable, in fact, that several early atmospheres and oceans were lost in massive impacts.[1]

Earth's earliest systematic diary entries overlap with the last pages of the Moon's, picking up again, after a 400-million year gap, about 4 billion years ago. Whorled metamorphic rocks exposed near the Great Slave Lake in northern Canada—the Acasta gneisses—are officially the oldest rocks on Earth (not merely mineral grains), and they mark the beginning of the Earth-based geologic timescale: the start of the Archean Eon. However, while the august Acasta rocks (and somewhat younger gneisses elsewhere in Canada, as well as Greenland and southern Minnesota) speak vividly of high-temperature upheavals deep inside the crust of the early Earth, they have no memories to share about conditions at the planet's surface.

The first rocks to provide glimpses of the light of day are the Isua *supracrustals* in southwestern Greenland, formed 3.8–3.7 billion years ago, at about the time the harsh fusillade of space debris was finally waning. The Isua sequence includes a variety of sedimentary rocks, which are records of erosion and deposition by surface water, as well as greenstones—metamorphosed but still recognizable "pillow" basalts, whose bulbous shapes are the signature of submarine eruptions. There were oceans on this ancient Earth, and the nearness of the Moon would have made tides significantly higher. Tides would also have been more frequent, because the day was significantly shorter, probably less than 18 hours (making a year of about 470 days).[2] Over time, friction between the ocean-atmosphere system and the solid Earth has acted like a soft brake that has gradually slowed the planet's rotation.

The Isua rocks provide indirect clues to Earth's second atmosphere. Their testimony to abundant water at Earth's surface 3.8 billion years ago would seem to be at odds with models of stellar evolution, which predict that our Sun, a *yellow dwarf* star, would have been about 30% less luminous than it is today. With so much less incoming solar energy, any water on Earth should have been frozen. This is the *faint young Sun* paradox first recognized by astrophysicist Carl Sagan in 1972.[3] Although there have been many creative proposals about how to reconcile this apparent contradiction between astrophysical theory and the rock record (with its echoes of earlier standoffs between physics and geology), the prevailing view is that an atmosphere dominated by greenhouse gases could have compensated for the dimmer Sun and made the early Earth's climate clement enough to keep ancient rivers rolling down to an open sea. Based on the atmospheres of neighboring Venus and Mars—the lingering breath of volcanoes—carbon dioxide ($CO_2$) and water vapor are likely to have been the primary heat-trapping gases, although methane, ethane, nitrogen, ammonia, and other compounds may also have acted as additional blankets that kept the Archean world warm. Whatever its exact greenhouse recipe, this second atmosphere would persist for more than a billion years, and would incubate the first Earthlings.

## SIGNS OF HABITATION

Their clearly aqueous origins make the Isua rocks an irresistible hunting ground for the spoor of early life. In 1996, a group of geologists from the United States, the United Kingdom, and Australia announced they had detected indirect geochemical evidence for life in graphite (carbon in mineral form) found in an iron-rich stratum at two outcrops of the Isua sequence.[4] In particular, they detected an unusual enrichment of the lighter-weight stable (i.e., nonradioactive) carbon isotope $^{12}C$ relative to the slightly heavier $^{13}C$. Carbon-fixing organisms, including photosynthesizing microbes and modern plants, are picky about their carbon. It takes slightly less energy to assimilate the lighter isotope, and so they will preferentially select it from the available pool of carbon atoms in their environments. Biogenic carbon thus has a lower $^{13}C/^{12}C$ ratio (by a few parts per thousand) than carbon that has not been processed by life-forms.

Like previous claims to the oldest evidence for life on Earth, however, this one was attacked on many fronts. Geologists from other research groups suggested, variously, that the rocks had been too metamorphosed to preserve the original carbon isotope signature;[5] that at one site, the host rock, which appeared to be a sedimentary formation, was in fact an igneous intrusion;[6] and that the samples had been contaminated by recent organic matter.[7] The number and vehemence of these critiques reflect the stakes involved: this is our Origin story.

As a result of these uncertainties, the prize for oldest documented evidence of life was provisionally returned to a rather similar, but 250 million-year-younger, sequence of greenstones and sedimentary rocks on the other side of the world. The Dresser Formation in the Warrawoona Group of northwestern Australia, after all, could boast direct, visible evidence for life: stromatolites (see figure 10).[8] These finely layered, lumpy rocks (the name means "mattress" or "quilt stone," a reference to their hummocky surfaces) are fossilized microbial mats that likely represent not just one species but a vertical ecosystem of prokaryotes living in symbiotic relationships in the primeval ocean. Sedimentary structures diagnostic of wave agitation indicate that stromatolites grew in shallow, sunlit waters and suggest that the organisms in at least their upper layers were photosynthesizers. Given their already sophisticated communal lifestyles, these stromatolite colonies cannot represent the very first life-forms; like the Jack Hills zircons, and Hutton's unconformity, they point backward to still-earlier unknown precursors. But for a time, Australia claimed not only the oldest surviving vestiges of crust but also the first traces of the biosphere.

Then in 2016, following two decades of discord about whether the Isua rocks contain the chemical ghosts of ancient organisms, a new group of geologists, including two authors of the original carbon isotope paper, published a new study documenting what appear to be plausible stromatolites in an outcrop of carbonate (limestone-like) rock at Isua, recently exposed as a result of the melting of an ice field.[9] Inevitably, much of the media coverage of this finding emphasized the implications for life on Mars, rather than the more salient point that life

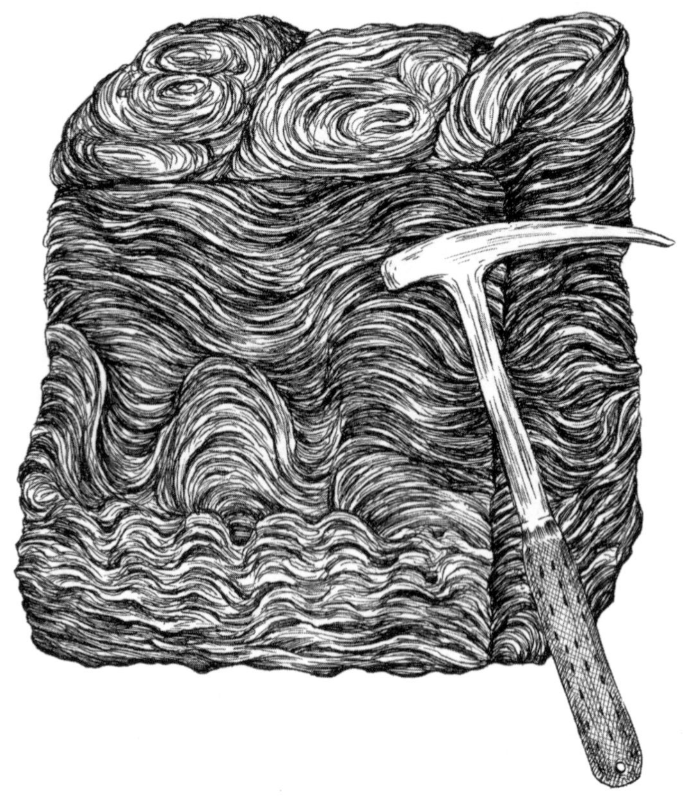

FIGURE 10. Stromatolites, fossilized (lower image), and alive and well at Shark Bay, Australia

on Earth seems to have appeared, and diversified, even while the planet was still being battered by extraterrestrial flotsam.[10] From this point on, the evolution of the air would be entangled with the saga of life on Earth.

## IRON AGE

Steel tycoon (and later, philanthropist) Andrew Carnegie—richer in his day than Bill Gates, Sam Walton, and Warren Buffett combined—amassed his fortune through the labor of thousands who toiled in his mills, but he actually owed everything to the work of ancient microbes. Carnegie's steel, and indeed almost all the steel ever produced in the world, was made with iron from a type of rock that is, in a sense, extinct. Most rock types—for example, the basalts that midocean ridge volcanoes exude, or sandstones composed of the granular remains of other rocks—are more or less timeless in the sense that they form on Earth today in the same way they have for billions of years. But the unimaginatively named sedimentary rocks called "iron formations" accumulated during only a specific interval in Earth's history and record a one-time revolution in the planet's surface chemistry in the Early Proterozoic Era, between about 2.5 billion and 1.8 billion years ago. In particular, these densest of rocks testify to changes in the air—the transition from a surface environment with no free oxygen ($O_2$) to a brave new world created by the rise of oxygen-emitting photosynthetic microorganisms like blue-green algae, or cyanobacteria (whose modern descendants are often called, with less than due respect, "pond scum"). This was Earth's third atmosphere.

The iron formations, found most notably in Australia, Brazil, Finland, and the Lake Superior region, are beautiful rocks with a striking color palette; fine laminae of silver hematite and black magnetite alternate with gray chert and red jasper. They can be many hundreds of feet thick, and are typically mined in giant open pits like the enormous "Hull Rust" chasm (the "Grand Canyon of the North") in Bob Dylan's hometown of Hibbing, Minnesota. Apart from their metallic composition, the iron formations have sedimentary characteristics very similar to those of modern limestones, indicating they must have been deposited in shallow marine environments. Yet in today's ocean, iron is in such short supply that it is a limiting nutrient—an essential element whose scarcity holds biological productivity in check. A controversial climate-engineering scheme is even based on this fact; the idea is that if the oceans were fertilized with iron powder, cyanobacteria would bloom and photosynthesize enthusiastically, and (if all were to go according to plan) sink to the ocean floor, sequestering large amounts of carbon, without (fingers crossed) wreaking havoc with the rest of the marine biosphere. In contrast with the trace amounts of iron in seawater today, the tremendous volume of the iron formations—visualize all the steel in the world's cars, aircraft, buildings, bridges, and railroads—attests to the great abundance of iron in the Proterozoic oceans.

It was oxygen, the insurgent gas first produced by cyanobacteria, that changed the rules about what could and could not be present in seawater. In the pre-oxygen regime, iron spewed by deep-sea volcanic vents was able to remain dissolved in the open ocean, commingling invisibly with sodium, calcium, and other ions. But when oxygen began to accumulate in shallow waters, it hunted down the iron atoms, bonded itself to them, and pulled them to the seafloor, creating iron formations. Oxygen purged the oceans of iron by literally rusting it out.

## A NEW WORLD ORDER

This geochemical coup d'état is known to geologists as the Great Oxidation Event, or GOE, and it represents a radical rewriting of the atmosphere-hydrosphere constitution. The presence of free oxygen changed the chemical interactions between rainwater and rocks on land, altering the composition of lakes, rivers, and groundwater. Certain types of cobbles that had been common in Archean river beds—particularly chunks of pyrite and uranium-rich minerals—disappeared from sedimentary deposits at this time, because they were now unstable or soluble under the new geochemical regulations. Conversely, modern oxide minerals—sulfates and phosphates like gypsum and apatite—became common entries in the rock record. Upstart life-forms had forced changes in the practices of the ancient mineral kingdom.

The presence of free oxygen ($O_2$) at Earth's surface also led to the establishment of an ozone ($O_3$) layer in the stratosphere, which shielded the surface environment from the ravages of ultraviolet radiation from the Sun and opened new frontiers to settlement. Novel alliances between oxygen and other elements made previously scarce nutrients such as nitrogen more mobile. This fueled major biological innovations, including more efficient photosynthesis, which produced even more oxygen. Like market opportunities created by a "disruptive" technological advance, entirely new biogeochemical cycles were established—global commodities exchanges mediated by single-celled organisms, through which large volumes of carbon, phosphorus, nitrogen, and sulfur were traded.[11] And in a strategic symbiotic merger, a tiny biological entrepreneur that had learned to process oxygen, called a mitochondrion, joined with a larger cell and founded the eukaryotic line that would eventually lead to plants and animals.

A continuing question about the GOE is why there was such a long lag between the first appearance of photosynthesizing life-forms 3.8 billion years ago and the emergence of free oxygen at about 2.5 billion years. One possibility is that the organisms that formed the stromatolites in the Isua and Warrawoona rocks performed anoxygenic (non-oxygen-producing) photosynthesis—a seeming oxy-moron (so to speak) to those of us familiar with plants—but a metabolic strategy that is still practiced by some bacteria that lurk in low-oxygen haunts like algae-clogged lakes. Rather than combining carbon dioxide ($CO_2$) and water ($H_2O$) in the presence of sunlight to form sugars $(CH_2O) \cdot n$ (where *n* is 3 or greater) and release oxygen ($O_2$), these microbes instead forge their sugar from $CO_2$ and hydrogen sulfide ($H_2S$, the "rotten egg" gas) and emit sulfur as the waste product.

Alternatively, it could be that the stromatolite-forming microbes did produce free oxygen, but that all of it was just as efficiently consumed when they decayed. Decomposition is the exact converse of photosynthesis—the same chemical reaction, but run in reverse: the sugars and other hydrogen-carbon compounds built by organisms react with free oxygen to yield carbon dioxide and water (burning hydrocarbons, a favorite human activity, is just a speeded-up version of this). So if photosynthesis and decay are perfectly balanced, there will be no net accumulation of $O_2$ in the air. This seems rather unlikely, however, to have been true for 1.3 billion years, given the tendency for at least some organic matter to be buried in sediments without decomposing (and eventually become those hydrocarbons we love to oxidize).

Another possibility is that for more than a billion years, any oxygen produced through photosynthesis quickly reacted with oxygen-hungry volcanic gases, especially hydrogen sulfide from seafloor volcanism. Then, around the end of the Archean, there may have been a transition to a more modern tectonic regime, with gases from subduction-related arc volcanism, which are less reducing, gaining in importance.[12] Some geologists, following the inborn urge for uniformitarianism, interpret Archean rocks like the Acasta gneisses and the Isua greenstones within the framework of present-day plate tectonics. A few uniformitarian zealots even argue for a modern-looking Earth back as far as Hadean time based on tenuous circumstantial evidence from the Jack Hills zircons. Others (full disclosure—I'm in this group) think we need to suppress Charles Lyell's voice in our head and consider the possibility of a different tectonic mode in Archean and Hadean times.

For one thing, the solid Earth was hotter (Lord Kelvin was partly right), and efficient subduction of ocean crust would have been unlikely. Also, while Archean rocks bear evidence of some sort of jostling and crumpling atop a convecting mantle, they don't have the same structural styles as those deformed at today's well-defined boundaries between rigid plates. Hotter, weaker slabs of crust might have piled up on each other and undergone partial melting, extracting the constituents to form granitic continents, leaving a deep layer of dense residual rock that sank back into the mantle in a process unappealingly called *drip tectonics*.[13] But starting with rocks from the end of Archean time, we can recognize the elements of modern crustal architecture: continental shelves, subduction zones, volcanic arcs, and full-fledged mountain belts that suggest Earth had cooled enough to form a brittle outer shell. So a nudge from a new tectonic system may have been enough to give oxygen production a slight lead over oxygen consumption. It seems entirely reasonable, in fact, that the Earth's tectonic coming of age would coincide with profound changes in the chemistry of the surface environment.

Although the Great Oxidation Event was a first-order disruption to the old geochemical establishment, its actual magnitude was not as great as its name suggests. Certain metallic trace elements in the banded iron formations such as chromium have stable isotopes whose behavior is highly sensitive to oxygen levels—Precambrian canaries, perhaps, in equally anachronistic coal mines. The ratios of these isotopes suggest that the Early Proterozoic atmosphere probably had only a small fraction—less than 0.1%—of the present level of oxygen (now 21% of the atmosphere by volume).[14] We Phanerozoic organisms would not have found this world hospitable. But the difference, in terms of chemical possibilities, between no free oxygen and even a little is greater than the difference between a little and a bit more.

## ONE BILLION YEARS OF LASSITUDE

After the upheavals of the GOE, Earth's atmosphere seems to have settled into a long period of geochemical stability. Although the main period of iron formation deposition ended around 1.8 billion years ago, oxygen levels seem to have remained about constant, and far below the current value, for another billion years after that.[15] Such sustained equilibrium—akin to a national economy that experiences no inflation, recessions, or market turmoil for decades—points to a remarkably fine-tuned balance between the oxygen supplied by hardy one-celled photosynthesizers and oxygen consumed by covetous metals, sulfurous volcanic gases, and decaying organic matter. This steady state may have been enforced by a regime of austerity—in particular, severe limitation on the availability of phosphorus, an essential nutrient for all life.

While shallow ocean waters had become oxygenated, there is evidence that deeper reaches remained in the transitional state of the Early Proterozoic. In these stratified conditions, precious phosphorus would have been continuously removed from deeper waters, stolen away on the surfaces of iron minerals, like currency smuggled out of a poor country in the linings of pilferers' coats. This in turn created chronic shortages of phosphorus in the shallow ocean. Biological productivity was thus kept in check, which limited organic carbon burial and in turn prevented atmospheric oxygen levels from rising.[16]

This lean eon encouraged organisms to pursue low-phosphorus lifestyles and new recycling strategies. In other ways, however, evolution seemed to be biding its time. The biosphere was diverse but still entirely unicellular; planktonic species—including some eukaryotic giants called *acritarchs*, up to 0.8 cm (0.3 in.) in diameter—proliferated in the oceans, and stromatolites quietly blanketed coastlines around the world. This peaceful stretch of the Proterozoic Eon has come to be known informally among geologists as the "Boring Billion." But this Homer Simpson–inspired designation is unfair, and misleading—akin to history books that focus only on war and skip over the much longer intervening periods of peace when "nothing happened."

First, maintaining such long-term equipoise is something that we humans in the Holocene might look to as a template for amending our own biogeochemical habits, since our looming environmental crises are the result of unchecked consumption of scarce resources and an extreme imbalance between the production and removal of an atmospheric gas. The Proterozoic Earth somehow "understood" the fundamental principles of sustainability; geochemical trading flourished, but all commodities flowed in closed loops—the waste products of one group of microbial manufacturers were the raw materials of another.

Second, the Boring Billion was the period when the durable cores of the modern continents were assembled, as the new

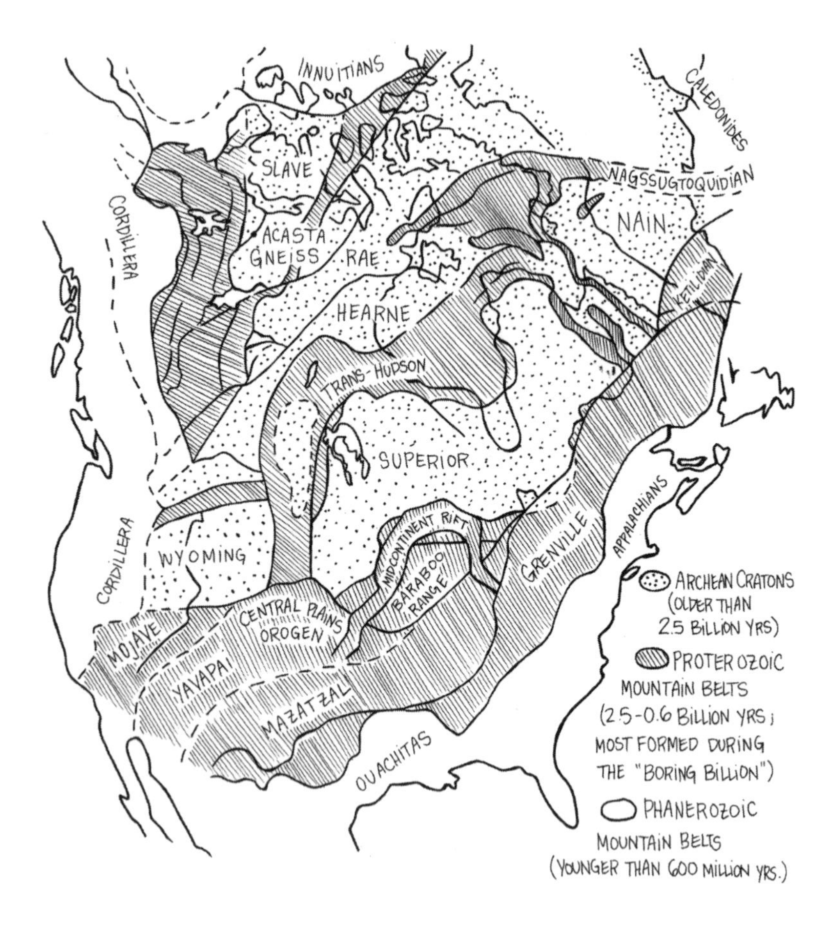


FIGURE 11. The United Plates of America—how North America was assembled

plate tectonic system swept together pieces of Archean crust and then constructed additions in the form of volcanic arcs. The basement rocks beneath my feet here in Wisconsin—and buried under younger sediments across the Midwest and Great Plains—are almost entirely Proterozoic, formed by mountain-building events during the Boring Billion, when vast areas of continental crust were annexed to the old Canadian Shield (see figure 11).

Boring, perhaps, but it was a productive time of infrastructure development—another practice we modern Earth-dwellers could profitably adopt.

Maybe because Proterozoic rocks and their stories are so familiar to me—the late great Penokee and Baraboo Ranges of the Lake Superior Region, the violent hotspot volcanoes of central Wisconsin, the immense Midcontinent Rift that nearly ripped North America apart—the Boring Billion doesn't seem that long ago. Thus it saddens me, irrationally, to know that in the equivalent amount of time into the future, about 1.5 billion years from now, the window of habitability will have closed for the Earth. The Sun, which is still getting brighter (at the very modest rate of about 0.9% per 100 million years), will have grown so luminous that the oceans will begin to vaporize, triggering a "moist greenhouse runaway."[17] Solar radiation will then break down water molecules into hydrogen and oxygen, which will be lost to space. In other words, if life first became viable after the early bombardment era ended 3.8 billion years ago, we are now almost three-quarters of the way through Earth's habitable period. Nevertheless, we should be grateful for the great wealth of time that this planet has had as a consequence of belonging to a yellow dwarf star with a lifetime of 10 billion years. Stars just 50% larger than the Sun have a life expectancy of only 3 billion years, which on Earth would be equivalent to the time span from the formation of the planet to the middle of the Boring Billion. At that point, Earth had so much more living to do.

## THE LONGEST WINTER

Things might have continued indefinitely in the monotonous Proterozoic mode, except that by around 800 million years ago, the new tectonic system had shepherded most of the continental crust into one large landmass that girdled the equator. Geologists call this ancient supercontinent Rodinia, from the Russian *ródina*, "motherland." Like all continents, Rodinia was only a temporary configuration, and it began to break apart through rifting by about 750 million years ago, creating expansive new coastlines at tropical latitudes. Rivers fed by heavy rains would have carried sediment and rock-derived elements to the sea, and organisms would have thrived in these comparatively nutrient-rich waters. High sedimentation rates on the continental shelves allowed organic carbon to be buried in significant volumes for the first time, which drew down atmospheric $CO_2$ levels and set Earth on a cooling trend.

Perennial sea ice would have begun to accumulate in the polar regions, increasing the *albedo*, or reflectivity, of the Earth's surface, which in turn led to further cooling—a classic example of positive feedback. Even as the ice advanced farther, carbon dioxide continued to be withdrawn from the atmosphere through both organic carbon burial and the intense chemical weathering of rocks on the low-latitude fragments of Rodinia (the mechanism by which the Himalaya drew down $CO_2$ and cooled the Earth in the Cenozoic). Once ice cover reached a critical point, the albedo effect would have led the planet into a "snowball" state—a perpetual snow day.

Exactly what happened during this Snowball Earth time—also called the *Cryogenian* Period, one of the few named divisions of the Proterozoic in common use—generates a lot of heat in the geologic literature. There is no disagreement that something went haywire for a time with the climate system. The rock record makes that clear: on almost every continent, rocks of this age are glacial deposits—either unsorted mixes of boulders and clay laid down directly by ice on land, or finely layered marine sediments punctuated by iceberg-rafted cobbles.

With much of the Earth's water locked up in glacial ice, sea level would have been lower by hundreds of feet, exposing large areas of the continents to erosion, at least until the deep ice age began and surface processes ground to a halt. The Great Unconformity in the Grand Canyon, between metamorphosed Proterozoic rocks like the Vishnu Schist and the first stratified unit, the Cambrian Tapeats Sandstone, is a record of Snowball Earth time in absentia. So while there is no question that an exceptional cold snap occurred at the end of the Proterozoic, specifics like how deep the freeze was, how the biosphere survived, and how the Earth emerged from its hypothermic state stoke the fires of academic debate.

## SPRING OF LIFE

But, clearly, the Earth did warm up again. Maybe the breath of volcanoes, which would have continued to erupt while other geologic processes had stopped, gradually coaxed Earth back from its cold coma over many thousands of years. Or perhaps a sudden, rude belch of long-sequestered biogenic methane from the seafloor transformed the icy planet into a hothouse in a matter of months or years. The resolution of the rock record and the precision of our dating methods are not fine enough for us to distinguish between these possibilities.

In any case, the end of Snowball Earth marks what could be called the Great Aeration, the second big step in free oxygen levels and the emergence of Earth's fourth, and current, atmosphere. Oxygen-sensitive trace elements in sedimentary rocks finally started behaving in the modern way, indicating that $O_2$ levels jumped from a fraction of a percent to something close to the present value. But the details of how the long-reigning quasi-oxygenated realm of the Proterozoic was overthrown are not known. Perhaps it was a large influx into the oceans of phosphorus, from rock powder ground up by Snowball glaciers, that kick-started marine life.[18] Or it might have been the energetic mixing of shallow and deep-ocean waters in the transition between ice-bound and greenhouse worlds that finally broke the geochemical stratification that had prevailed for 1.5 billion years.

Once oxygen levels rose even a bit higher, organisms that evolved to use it in their metabolic processes were significantly more efficient in extracting energy from the environment and were able to grow larger than any had before. Within a million years of the end of Snowball Earth, a strange new macroscopic ecosystem of puffy organisms called the *Ediacaran fauna* appears in the fossil record at sites around the world, including southern Australia, the White Sea region of Russia, Leicestershire in England, and Newfoundland in Canada. These bizarre down-parka-like organisms were shaped like Frisbees and fern fronds, the latter up to 1 m (3 ft) high, with holdfasts to anchor them to the seafloor. They had neither guts nor mineralized shells, suggesting their world was a peaceable kingdom of osmotic nutrition without threat of predation. Some may have been precursors to later, more familiar marine lineages such as the brachiopods, or lampshells. But others seem to have been early evolutionary experiments in building bigger life-forms that left no modern descendants.

The Ediacarans' moment in the avant-garde was brief, however. Within about 40 million years, the seafloor had become the venue for the period of frenzied anatomical tinkering called the Cambrian explosion, when the first carnivores set off an arms race between predator and prey. Like Wile E. Coyote and the Road Runner, they've been trying to outwit each other ever since. Hard protective shells of calcium carbonate became obligatory for bite-sized organisms; specialized swimming gear and killing apparatuses de rigueur for the big meat eaters.

The pace of evolution in the Cambrian explosion continues to be a topic of some controversy, pitting paleontologists against biologists who use genomic approaches to determine when different branches of the tree of life first emerged. The fossil record suggests that the interval between about 540 and 520 million years ago was a time of unprecedented, and never to be repeated, biological innovation. But this is at odds with various *molecular clocks*, which are based on the assumption that protein-coding genes accumulate substitutions at a constant rate in evolutionary lineages. Most of the molecular analyses suggest that Kingdom Animalia, whose first members were probably sponges, was founded in the late Proterozoic, 750–800 million years ago and that the Cambrian "Explosion" may instead have been a slow-burning fuse.[19] This, however, puts our infancy in the bleak and frigid time of Snowball Earth, which would seem an unlikely nursery. The disagreement reveals interesting cultural differences between field-based paleontologists, who, inured to the idiosyncrasies of fossil life, are willing to embrace the idea of nonsteady rates of evolution, versus lab-based molecular biologists, who see mechanism in cellular structure and are more orthodox uniformitarians than their geologic counterparts. While the Precambrian is by no means the obscure, unmapped expanse it was to Victorian geologists, the transition from it across the threshold into the Cambrian remains dimly lit.

In most paleontologic textbooks, the Cambrian explosion is the start of the story, the prelude to the rollicking tale of trilobites, lungfish, coal swamps, tyrannosaurs, pterodactyls, megatheria, mammoths, and hominids. In the most important ways, though, the Cambrian world was not so different from the modern biosphere—almost all the main animal phyla were already present, and for the next 500 million years those same players would organize themselves into elaborate oxygen-dependent ecosystems with multitiered food webs, expanding onto the continents and into the skies, developing ever more specialized adaptations to their ambient environments. And for the next 500 million years, they suffered spectacularly whenever those environments, and especially the atmosphere, changed too fast.

## CURTAINS

In the nineteenth century, the field of geology was primarily devoted to paleontology, and even before Darwin's *On the Origin of Species* was published in 1859, fossils were being used to demarcate divisions of geologic time. Victorian geologists studiously catalogued gradational changes in certain lineages like the coiled ammonites, whose shells bear ornate patterns as distinctive to particular moments in time as hoopskirts or saddle shoes. But geologists also recognized points in the rock record at which the changes in the fossils were not merely incremental alterations in costume detail but wholesale replacement of one cast of characters with an entirely new troupe. On the basis of such discontinuities, John Phillips—the nephew of canal digger William Smith, who had developed the concept of index fossils—proposed in 1841 that there had been three great chapters in the history of life: the Paleozoic, Mesozoic, and Cenozoic (Old, Middle, and Recent Life) eras. (The much deeper roots of life in the Archean, more than 3 billion years before the start of the Paleozoic, would not be appreciated for another century).

Phillips, an orphan, was raised by William Smith and accompanied him as a child on many fossil expeditions. He was an excellent and perceptive paleontologist but became a vocal opponent of Darwin's theory of evolution by natural selection, believing instead that the exquisite match between animals and their environments was evidence of a divine plan (which apparently allowed do-overs). In the later part of his career, Phillips allied himself with Lord Kelvin to undermine Darwin's assertions about the "prodigious durations of the geological epochs."[20] Still, his chapter designations for the epic story of animal evolution were astute.

Darwin was understandably irritated by Phillips but could not deny that the fossil record did seem to have some sudden and perplexing disappearances. Confident that evolution proceeded at a consistent pace, however, he did not perceive these as evidence for natural catastrophes. Darwin fully accepted the concept of extinction; indeed, the continual culling of organisms was central to his theory. But he argued that what appear in sedimentary rock sequences to be sudden extinction events were simply artifacts of the intermittent nature of sedimentation. He devoted an entire chapter in *Origin* to "The Imperfection of the Geologic Record," in which he emphasized that rocks document only a fraction of elapsed time, stating, "between each successive formation, we have, in the opinion of most geologists, enormously long blank periods." Darwin also suggested that the rates of sedimentation, when it does occur, may not be fast enough to capture evolution in progress: "Although each formation may mark a very long lapse of years, each perhaps is short compared with the period requisite to change one species into another." He further speculated, perceptively, that our reading of the fossil record is skewed by the fact that we can find fossils only in settings where sediments once accumulated (otherwise there is no rock), but those settings are not always the places where organisms may have lived.

Darwin's inclination to explain away discontinuities in the fossil record would prevail well into the mid-twentieth century, when the geologic timescale was well enough calibrated to make it undeniable that on occasion bad things had suddenly happened to good ecosystems. We know now that there have been at least five great *mass extinctions*, and many smaller ones, since the start of Cambrian time. After each of these, life on Earth eventually recovered but was irrevocably changed, with the creatures that survived, as much by happenstance as hard-earned fitness, becoming the unlikely founders of brave new biospheres.

## APOCALYPSE NOW

In a mass extinction, the normally meticulous scalpel of natural selection, which excises this moth or spares that finch on account of the tiniest differences in wing color or beak shape, becomes the evolutionary equivalent of a machete. Whole taxonomic groups of organisms—not merely individuals or species but genera, families, and orders—in many locations and habitats are cut down in swift, indiscriminate strokes. The cause of a mass extinction is generally very different from the factors behind ordinary thinning by natural selection, in the same way that deaths from wars or epidemics differ in a fundamental way from deaths due to individual accidents or illness. Paleontologists quantify the severity in terms of the magnitude of deviation from the background rate of extinction for different groups. The background rate of extinction for amphibians in the Cenozoic, for example, is less than 0.01 species/year or about one frog or salamander per century.[21] Mass extinctions imply that the normally commensurate tempos of evolution and environmental change—well matched over time, in the same way that tectonics and erosion keep pace with each other—have fallen out of synchrony. Gradual geologic change—the growth of mountain belts, the separation of continents—inspires the biosphere to innovate, but abrupt shifts may devastate it. In mass extinctions, alterations to the environment have for some reason accelerated to the point at which much of the biosphere cannot keep up.

It is fascinating to look back at hypotheses for the end-Cretaceous extinction described in the textbook for the Earth History course I took in college in the early 1980s, just before the Alvarez meteorite impact hypothesis began to gain traction in the geologic community. Old evolutionarily untenable ideas about the dinosaurs being sluggish and stupid—and by implication "deserving" of extinction—had by then given way to new depictions of creatures that were sprightly, warm-blooded, sociable (in some cases), and even smart. So killing them off had become harder, and none of the proposals about their demise—global cooling, virulent plagues, genocide by egg-eating mammals, deadly allergies to the just-evolved flowering plants (!)—seemed to be shocks sufficiently short and sharp to do the job. The single extraterrestrial hypothesis mentioned was the notion that cosmic radiation from a distant supernova might have reached Earth just at the moment when there was a magnetic field reversal and the planet was least protected—a literal *disaster* in the Greek meaning of the word: "bad star."

Reading these ideas now feels like revisiting a kinder, gentler moment in history, because scientific ideas about mass extinctions seem to parallel contemporary sources of existential angst in society; the geologic past often acts as a screen onto which we project our deepest fears. This is not to say that hypotheses about mass extinctions are unscientific, but that terror of new types of apocalypse helps fuel our imaginations about possible scenarios for cataclysms of the past. Geologists, as humans who live in particular social settings and historical moments, cannot help but be influenced by the prevailing zeitgeist. Compared with the jittery angst of the twentieth and twenty-first centuries, the Victorian period was a time of great optimism about the potential of technological and scientific progress to improve the lot of mankind. So besides the Lyellian taboo against invoking geologic catastrophes (specifically the old-fashioned biblical type), it may be that because the Victorians were not haunted by visions of the end times, Armageddon was simply not in the scientific air.

In 1980, however, fearsome technological advances that the Victorians could not have foreseen threatened human civilization, and it was at that anxious moment late in the Cold War that the Alvarez meteorite impact hypothesis emerged. Its description of a dusty shroud of pulverized rock blasted into the stratosphere, blocking photosynthesis and leading to mass starvation, came directly from the "nuclear winter" scenarios of Carl Sagan and atmospheric chemist Paul Crutzen in the 1970s. The eruption of Mount Saint Helens that same year made it even easier to imagine an ashy doomsday.

By the time the Chicxulub crater was identified in 1990, the Berlin Wall had fallen. As the threat of nuclear holocaust began to fade from the collective consciousness, it was replaced by a growing awareness that environmental malefactions might be humanity's downfall. Acid rain was shown to be devastating forests in New England and Scandinavia, the legacy of sulfurous emissions from decades of coal burning. The selective pattern of marine extinction at the end of the Cretaceous, with shelled creatures in deep water faring better than those in the shallows, suddenly looked very much like what one would expect in an ocean that had become soured by sulfuric acid. And the rocks in the Yucatán crater had plenty of sulfur in them: they included thick layers of a mineral called anhydrite, or anhydrous calcium sulfate, which would have been vaporized in the impact, hurled into the atmosphere, and then precipitated as burning acid rain. The 1991 eruption of Mount Pinatubo, in the Philippines—10 times more powerful than that of Mount Saint Helens—provided further insight. The eruption injected enough sulfate particles into the stratosphere to counteract, for two years, the inexorable climb in global temperatures related to rising greenhouse-gas concentrations. The immense volumes of brimstone blasted from the 240-km (150 mi)-wide Yucatán crater could have caused far more severe cooling—devastating to organisms accustomed to the warm Cretaceous world—before falling out of the atmosphere as the rain from hell. It seemed, then, that sulfur, not just dust, must have been the real culprit in the end-Cretaceous extinction.

But many paleontologists remained unsatisfied with this explanation, too. Caustic acid rain should have been especially harmful to freshwater ecosystems, yet species in these environments, including frogs and other amphibians sensitive to changes in water chemistry, had survival rates of close to 90%—far higher than those that lived on dry land, where only 12% withstood the cataclysm. The failure of any of the proposed kill mechanisms to account for the details of the fossil record has led some paleontologists to propose that the asteroid was not a lone assassin but struck a global ecosystem already weakened by other injuries. The most frequently cited accomplice is volcanic activity, in particular the eruptions that produced the Deccan Traps, a 1.6 km (1 mi)-thick stack of basalt flows in present-day India. For tens of thousands of years leading up to the extinction, the oozing lavas released enormous quantities of carbon dioxide, creating a world that was already in environmental peril when it was mortally injured by a blow from space. Vaporization of a thick sequence of limestone at the Chicxulub site would have injected even more $CO_2$ into the air, so that after a few years of frigid cold from the pall of ash, the climate whipsawed into a withering hothouse. In recent reconstructions of the Cretaceous finale, the murderous but charismatic asteroid has been forced to share the stage with far less glamorous greenhouse gases.

## BAD AIR DAYS

The study of mass extinctions became a distinct and fashionable subdiscipline within paleontology in the decade after the end-Cretaceous impact was proposed. To those who embraced the newly "legalized" catastrophism, it seemed likely that all mass extinctions could eventually be blamed on extraterrestrial impacts. A brilliant paleontologist Jack Sepkoski of the University of Chicago, who was the first to recognize the potential of Big Data in paleontology, believed he had detected a 26-million-year cycle in extinction frequency through an analysis of thousands of fossil catalogs. In a strange kind of neo-uniformitarianism, he speculated that episodic die-offs might be linked with Earth's periodic passage through the spiral arms of the galaxy, which could destabilize the orbits of comets.[22] This inspired eager searches for evidence of large impacts at the times of other mass extinctions, and moved the study of impact cratering from a fringe field into the geologic mainstream. But three decades later, no other major biological crisis has been convincingly linked with the crash landing of a comet or asteroid. We are left with the sobering fact that sometimes things can go horribly wrong for life on this planet, for reasons completely internal to the Earth system.

Besides the end-Cretaceous cataclysm, the other great extinctions include, chronologically (1) the Late Ordovician event about 440 million years ago, which was the first major pruning following the Cambrian explosion; (2) a closely spaced pair of die-offs in the late Devonian Period (about 365 million years ago), by which time macroscopic life had moved onto land; (3) the end-Permian holocaust 250 million years ago, the mother of all mass extinctions, which John Phillips aptly marked as the close of the Paleozoic Era; and (4) the Late Triassic event, a cruel blow just 50 million years after the Permian debacle. Depending on how one measures the severity of these massacres (by numbers of species or genera or families vanquished), the dinosaur extinction is the fourth or fifth in rank.

Although the victims and the circumstances of these calamities differ in detail, they share some striking similarities (appendix III). All—including the end-Cretaceous event—involved abrupt climate change, and all, with the exception of the Devonian event (when tropical seas turned cold), are linked with rapid warming. Second, all involved major perturbations to the carbon cycle and carbon content of the atmosphere, either by unusually effusive volcanism (Permian, Triassic, Cretaceous) and/or through an imbalance between carbon sequestered by the biosphere and carbon released from stored hydrocarbons (Ordovician, Devonian, Permian, Triassic). Third, all entailed rapid changes in ocean chemistry, including acidification that devastated calcite-secreting organisms (Permian, Triassic, Cretaceous) and/or widespread anoxia (*dead zones*), which asphyxiated almost everybody except for sulfur-loving bacteria (Ordovician, Devonian, Permian). All the extinctions, in fact, were followed by a period of time—hundreds of thousands to millions of years—when microbes alone thrived while the rest of the biosphere struggled to get back on its feet (or into its shells). The great mass extinctions challenge any conceit that we are the triumphant culmination of 3.5 billion years of evolution. Life is endlessly inventive, always tinkering and experimenting, but not with a particular notion of progress. For us mammals, the Cretaceous extinction was the lucky break that cleared the way for a golden age, but if the story of the biosphere were written from the perspective of prokaryotic rather than macroscopic life, the extinctions would hardly register. Even today, prokaryotes (bacteria and archea) make up at least 50% of all biomass on Earth.[23] One might say that Earth's biosphere is, and always has been, a "microcracy," ruled by the tiny. When larger, arriviste life-forms falter, infinitely adaptable microbes, whose evolutionary timescales are measured in months rather than millennia, are always eager to move in and reassert their long-held dominion over the planet.

Perhaps most importantly, none of the mass extinctions—even the relatively "clean" Cretaceous disaster—can be fully attributed to a single cause; all involved rapid changes in several geologic systems at one time, which in turn triggered knock-on effects in still others. In some respects, this is reassuring; it means that it takes a "perfect storm" of convergent causes to destabilize the biosphere. Nevertheless, many of the malefactors—greenhouse gases, carbon-cycle disturbances, ocean acidification, and anoxia—are uncomfortably familiar to current residents of Earth. And if a looming catastrophe has multiple origins, there will be no precise predictions and no silver-bullet solutions.

The story of the atmosphere reminds us that the sky over our head is not the only, or ultimate, one to shelter the Earth. When there is change in the air, even after long periods of stability, it can blow through with breathtaking suddenness, as Svalbard's withering glaciers attest. In the aftermath of these winds of change, upheavals in biogeochemical cycles ripple through ecosystems at all levels. Organisms that have invested everything in the old world order will suffer or even be extinguished while microbes quietly clean up the mess and decree a new set of rules for the survivors. Tinkering with atmospheric chemistry is a dangerous business; ungovernable forces can come out of thin air.

CHAPTER 5

# GREAT ACCELERATIONS

Any fool can destroy trees; they cannot run away.

—JOHN MUIR, *OUR NATIONAL PARKS*, 1901

## ACCIDENTAL VANDALS

At most U.S. colleges and universities, earning a degree in geology requires completion of a rite of passage called "field camp." Traditionally, this is a six-week course in a Western state with rugged topography and plenty of bare rock baking in the sun. Aspiring geologists learn to map rock units and mineral deposits, log stratigraphic sequences, draw cross sections, and interpret landforms. In the old days, field camp was the course that separated "the men from the boys." Fortunately, my own field camp at the University of Minnesota was taught by professors with a more enlightened philosophy. Even though Minnesota has plenty of interesting rocks of its own, our field camp was set in the spectacular Sawatch Range of central Colorado.

We had a day off each week, and during one of those times of sweet liberty, a group of us set out on a long hike to explore an old pegmatite mine we had heard about. Pegmatites are exotic igneous rocks famed for their oversized crystals of rare and colorful minerals and valued, increasingly, as sources of rare-earth elements, which are essential for high-tech batteries, cellphones, and digital storage media. Pegmatites represent the very last stage of solidification of some granitic magmas, when a combination of undercooling and a high content of magmatic gases allows crystals to grow many times faster than usual. A normal quartz or feldspar crystal forming in a magma chamber beneath a volcano like Mount Saint Helens might grow at the leisurely pace of about 0.6 cm (0.25 in.) per century.[1] Pegmatite crystals, on the other hand, are the baby blue whales of the mineral kingdom, bulking up at the dizzying rate of inches per year.[2] Although they can form quickly under the right circumstances, pegmatites are rare—not exactly renewable resources. The pegmatite we were hunting was an old one, formed in Mesoproterozoic time at least 1.5 billion years ago, long before the modern Rockies existed.

We found the road into the abandoned mine diggings—hesitating briefly at the "No Trespassing" signs—and followed a string of waste rock piles to a hollowed-out space on the side of a half-blasted hill. There we discovered what pegmatite zealots (a distinct subculture of mineralogists) call a *gem pocket*. It was like stepping into the pastel world of an old-fashioned Easter sugar egg: giant crystals of white feldspar were decorated with clusters of purple mica (lepidolite) and hexagonal prisms of pink and green tourmaline. Some of the tourmalines were perfect gemstone miniatures of watermelon slices, with thin green rinds and pink interiors. In an instant, we were all seized with a visceral greed, a need to take as many of these treasures as we could. We had come with our rock hammers, but the pick ends were blunt, designed for breaking rocks, not extracting delicate crystals. I managed to tap out a few small deep-pink tourmalines, and then spotted a prize: a perfect watermelon-colored crystal about 8 cm (3 in.) long. It was in an awkward corner close to the ceiling of the excavation, with little room for wielding a hammer, but I was determined to have it. I began pounding away, thinking ahead to how I would display this trophy at home when, in one errant blow, I smashed it.

In that moment it seemed my vision was suddenly cleared, as if I had been released from a malevolent spell that had engulfed us when we entered the gem pocket. I abruptly lost my appetite for the whole enterprise. After several years of immersion in the world of geology, I had developed some sense for Deep Time. And I saw that in an avaricious second I had carelessly destroyed an exquisite thing that had witnessed a third of Earth's history—most of the Boring Billion, Snowball Earth, the emergence of animals, the great extinctions, the growth of the Rockies. I felt sickened by the scene of devastation around me, and my complicity in it.

I have the same feeling now in watching the demise of Svalbard's glaciers—and the increasingly anemic winters we have in Wisconsin—knowing that I am culpable for them as a person who loves international travel and long hot showers, and more generally as a member of a fossil fuel–addicted society. In my lifetime, we have thoughtlessly smashed ancient ecosystems and made a wreckage of long-evolving biogeochemical cycles. We have set in motion changes for which there are few precedents in the geologic past and which will cast long shadows far into the future.

## AN ANTHROPOCENE ALMANAC

Sometime in the last century we crossed a tipping point at which rates of environmental change caused by humans outstripped those by many natural geologic and biological processes. That threshold marked the start of a proposed new epoch in the geologic timescale, the Anthropocene. The term was coined in 2002 by Paul Crutzen, a Nobel Prize–winning atmospheric chemist, and it quickly entered both the geologic literature and popular lexicon as shorthand for this unprecedented time when the behavior of the planet bears the unmistakable imprint of human activity.

In 2008, a short paper by a group of stratigraphers in the Geological Society of London provided quantitative arguments for how the Anthropocene could be formally defined.[3] The authors pointed to five distinct systems in which human activities have at least doubled the rates of geologic processes. These include the following:

- erosion and sedimentation, in which humans outpace all the world's rivers by an order of magnitude (a factor of 10);
- sea level rise, which had been close to nil for the past 7,000 years[4] but is now about 0.3 m (1 ft) per century and expected to be twice that by 2100;
- ocean chemistry, also stable for many millennia but now 0.1 pH unit more acidic than a century ago;
- extinction rates, now a factor of 1000 to 10,000 above background rates;[5]

and of course

- atmospheric carbon dioxide, which at more than 400 ppm is higher than at any time in the last 4 million years (before the Ice Age), while emissions by human activities surpass all those of the world's volcanoes by a factor of 100.[6]

Other authors note that phosphorus and nitrogen efflux into lakes and coastal waters—leading to anoxic dead zones—is now more than double the natural rates, due to runoff of agricultural fertilizers.[7] And through agriculture, deforestation, fires, and other land-use practices, humans dictate one-quarter of the *net primary productivity*—the total photosynthetic effort of plants—on land.[8]

Most geologists think these stark facts more than justify the adoption of the Anthropocene, not only as a useful concept but also as formal division of the geologic timescale, on par with the Pleistocene (the Ice Age, 2.6 million to 11,700 years ago) and the Holocene (essentially, recorded human history). The magnitudes of human-induced planetary changes, "achieved" in less than a century, are equivalent to those in the great mass extinctions that define other boundaries in geologic time. With the exception of the end-Cretaceous meteorite impact, however, those events unfolded over tens of thousands of years.

The International Commission on Stratigraphy—that formidable parliament of Time—has taken the matter up, and the main disagreements are bureaucratic: in particular, how exactly to define the start of the Anthropocene. Should there be a Global Boundary Stratigraphic Section and Point (GSSP, or "golden spike") as for other boundaries in geologic time? The GSSP for the base of the Holocene is a particular layer within the Greenland ice cap, with an isotopic signal that marks the onset of the warmer climate of the Holocene.[9] Ice is more ephemeral than rock, but the layer lies more than 1400 m (4600 ft) below the surface and is safe from melting for now. (Also, a sample of the layer is archived in freezers at the University of Copenhagen). The Anthropocene could be similarly defined by a distinctive signature in polar ice—perhaps the spike in unusual isotopes that is the shameful legacy, the Scarlet *A*, of atomic bomb tests in the 1950s and '60s. But this near-surface ice could very soon be a victim of Anthropocene climate; glacial archives are being lost at an alarming rate around the world. On the Quelccaya Ice Cap in the High Andes, for example, 1600 years of ice has vanished in the past two decades—destroying a high-resolution weather record going back to the time of the Nazca people.[10] The use of the word *glacial* to mean "imperceptibly slow" is quickly becoming an anachronism; today, glaciers are among the rapidly changing entities in Nature.

Some geologists therefore suggest that an exception be made in defining the Anthropocene and that a calendar year—perhaps 1950—rather than a natural archive be chosen as its formal beginning. After all, we humans are the only ones agonizing over this, and as long as we're around we can remind each other of the date. If at some point we vanish, it is likely no one else will fret about the definition of the Anthropocene. In many ways, the exact start of the Anthropocene matters less than the concept behind it.

A subtler point for geologists is that the idea of the Anthropocene represents a fundamental break with the philosophical underpinnings of the field, established by Hutton and Lyell. Hutton's great insight was that the past and present are not disjunct domains governed by different rules but linked through the continuity of geologic processes. And much of Lyell's magnum opus, *Principles of Geology*, is a polemic intended to dissuade readers from the idea that geologic change happened faster in the past than in the present. The Anthropocene now inverts this idea by emphasizing how processes are faster in the present than in the past. In attempting to predict the geologic future without the comfort of uniformitarianism, we are in a position strangely analogous to that of pre-nineteenth century geologists who had no guidelines for understanding the geologic past. Still, we can refer only to the recent geologic record for possible analogs to our present uncertain moment in time.

## UNDER THE WEATHER

Climate-controlled buildings and the year-round availability of fresh fruit allow citizens of wealthy nations in the twenty-first century to treat the weather as a backdrop to their lives, not the main story. We may complain about the inaccuracy of a local forecast or be irritated when rain foils weekend plans, but as a society, we largely ignore the weather until it disrupts everyday life. Rather than measure the value of good weather (imagine this headline: "Last Week's Sunshine Was Worth $10 million to Area Farmers"), we characterize bad weather events—blizzards, hurricanes, heat waves, droughts, floods—as costly anomalies that deprive businesses of their "rightful" earnings. That is, we assume that the weather is normally stable and benign, and are constantly surprised when it is not.

The long-term imprint of weather and climate on human civilization is the focus of a new area of interdisciplinary scholarship that integrates history, economics, sociology, anthropology, statistics, and climate science. One of the salient patterns that emerges when one looks at the past two millennia of human civilization is that periods of social instability and conflict coincide, at a high level of statistical significance, with intervals when climate deviates even modestly from normal.[11] In early medieval Europe, for example, average temperatures only one degree lower than average led to crop failures and spurred the great mass migrations and intertribal clashes of the period from AD 400 to 700. Sustained drought related to changes in Pacific Ocean weather patterns around the year 900 caused the collapse of both the Mayan civilization in Central America and the Tang dynasty in China. The Angkor kingdom of Southeast Asia, which had flourished for 500 years, crumbled after just two decades of drought in the early fifteenth century. Another cold period in Europe coincided with the Thirty Years War, from 1618 to 1648, which was more devastating even than World War I in terms of the percentage of the population killed. Although the war was nominally a religious and political conflict, the animosities were deepened and the suffering exacerbated by climate-related famine.

We may think that in modern times we are no longer so vulnerable to mere weather phenomena. But analysis of global police records from the past half-century shows that for each standard deviation increase in average temperature in major cities, violent crime rates rose by 4%. A similar statistical study finds that climate stresses like water shortages have caused local and regional intergroup conflicts around the world to increase by at least 14% in recent decades.[12] And in many ways, our advanced technologies make us less flexible than previous societies in the face of change. We have made huge infrastructural investments in coastal cities based on a bet that sea level will remain constant. We have built sprawling cities in the desert on the assumption that snow and rain will keep refilling reservoirs. We have a food production system that is predicated on the belief that old, familiar weather patterns will always return.

But the weather is getting weird. Ten of the hottest years on record have occurred since the start of this millennium. "'One hundred-year"' and "500-year" flood events are happening once a decade. The new rules of the Anthropocene are even making it difficult for Earth scientists to use the quantitative models they have developed to study geologic systems. Such models are based on the concept of *stationarity*—the idea that natural systems vary within a well-defined range with unchanging upper and lower bounds, an assumption that has yielded reasonable predictions in the past. A sobering report by an international group of leading hydrologists recently stated that "stationarity is dead and should no longer serve as a central, default assumption in water-resource risk assessment and planning."[13] In other words, the main prediction about weather and the water cycle is that they will become increasingly unpredictable.

Yet the public clings to an optimistic belief in uniformitarianism. This is partly understandable, because it is rooted in the geologic fact that climate in the Holocene Epoch, which saw the rise of everything we associate with human civilization—agriculture, written language, science, technology, government, fine arts—has been exceptionally stable. In fact, this stability is arguably the very thing that allowed humans to build civilizations at all. The large-amplitude climate oscillations of the Pleistocene, in contrast, probably kept nascent human societies in check. The "Ice Age" wasn't, in fact, entirely icy; instead, for 2.5 million years, the climate fluctuated manically over many timescales—as glaciologist Richard Alley memorably puts it, like someone "playing with a yoyo while bungee jumping off a roller coaster."[14] Understanding what exactly was going on in the Pleistocene is essential for putting current rates of climate change in perspective. The story of deciphering the Ice Age takes us back once again to Lyell, but also involves Swiss farmers, a Scottish janitor, and a Serbian mathematician.

## WARMING UP TO ICE

Here in Wisconsin, large boulders of granite and gneiss are common centerpieces of upscale landscaping around medical complexes and office buildings. In the early nineteenth century, such stones—often completely different in composition from the local bedrock—were among the most vexing mysteries facing geologists in the Great Lakes states and northern Europe. These *erratics*, scattered far from their sources, seemed to support the biblical idea of a Great Flood. Consequently, the rocks, and the clayey deposits they were often lodged in, became known as *diluvium* (sediment left by the deluge) or *drift* (a rather gentle term, considering the force of water that would have been necessary to transport such material). The latter term persists, anachronistically, in the name used for the distinctive region of deep bedrock valleys in southwestern Wisconsin—the "Driftless Area," where no erratics or other types of diluvium are found.

A Swiss geologist, Louis Agassiz (1807–1873), is usually credited as the first to propose, in 1838, that great ice sheets, not floodwaters, might have carried erratic boulders long distances. Agassiz is championed in geology textbooks as a revolutionary thinker, but it seems a German naturalist, Karl Schimper—who in fact coined the term *Eiszeit* ("Ice Age")—had earlier reached the same conclusion and shared it with Agassiz on a joint outing in the Alps.[15] Schimper's insights, in turn, may have come from Swiss farmers, who understood glaciers and to whom it was obvious that large boulders strewn far down alpine valleys marked the former positions of ice masses. More unforgivably, Agassiz later used his scientific credentials and his position as a Harvard professor to advance completely unscientific and abhorrently racist theories of human evolution; in my view, he should have an asterisk next to his name in the annals of science, like an athlete whose medal was rescinded for doping. Unfortunately, he remains the eponym of a giant late-Pleistocene lake, glacial Lake Agassiz, that covered much of North Dakota, Minnesota, and Manitoba (and left them so famously flat).

While Charles Lyell disavowed the Flood, he also disliked the idea of an Ice Age during which large areas of now-temperate Europe and North America had been covered by ice. If not exactly catastrophic, it was certainly non-uniformitarian. But as geologists began to map the patterns of "drift," the idea of a great Ice Age was shown to have explanatory power. In the upper Great Lakes region, it became clear that there had in fact been not one but several ice advances and retreats, each leaving distinct deposits (though each, strangely, avoiding the Driftless Area). What could be causing such cycles of warming and cooling?

As early as the mid-nineteenth century, some scientists began to explore the hypothesis that variations in Earth's orbital habits could affect the way sunlight falls on the Earth and potentially trigger episodic ice ages. The gravitational influences of the Moon and neighboring planets cause cyclical changes in three aspects of Earth's motion in space: (1) the elliptical shape, or eccentricity, of the Earth's orbit around the Sun, which stretches and shrinks on a 100,000-year timescale; (2) the tilt, or obliquity of Earth's rotation axis, which varies between about 21.5° and 24.5° every 41,000 years; and (3) the slow wobble, or *precession* of the planet, like a toy top, which changes the hemisphere that is pointed toward the Sun at the solstices over a cycle that averages 23,000 years. Today, these three variables are called *Milankovitch cycles* for the alliteratively named mathematician Milutin Milankovitch (1879–1958) who, in spite of his status as a displaced person for most of two world wars, managed to work out the combined effects of these cycles on solar irradiance of Earth.

But it was actually a self-educated Scotsman, James Croll (1821–1890), who had performed the first arduous calculations of the orbital cycles, more than 50 years earlier (a fact Milankovitch fully acknowledged). Croll had a keen mathematical mind and great interest in science but was too poor to attend even secondary school. After some years as an innkeeper, he took a job as a janitor at Anderson College in Glasgow, where he would study scientific volumes in the library late at night (in a nineteenth-century real-life version of the plot of the 1997 film *Good Will Hunting*). In the 1860s, he began a correspondence with Charles Lyell about his calculations of orbital variations and their effects on climate. Lyell, who by this time had reluctantly accepted glacial theory, was impressed by Croll's clear brilliance and helped him gain a position at the Geological Survey of Scotland. (Croll also exchanged letters with Darwin on the question of erosion rates). Croll's work seemed to suggest that ice ages would be out of synchrony in the Northern versus Southern Hemispheres owing to the opposite effects of precession in the two regions. This reasoning appealed to Lyell, since it meant that on average, the Earth maintained a steady state, an idea Lyell could not relinquish. A half-century later, Milankovitch would recognize that because of the disproportionate concentration of landmasses in the Northern Hemisphere, the influence of precessional cycles on northern latitudes actually dominated global climate.

Neither Croll nor Milankovitch had any high-resolution geologic data against which to test their calculations, however. By the 1880s, the eminent Wisconsin geologist T. C. Chamberlin (for whom my once-glaciated Svalbard valley was named) had documented four distinct glacial periods, which he named for the states in which their deposits were best preserved—starting from the most recent, the Wisconsinan, Illinoisan, Kansan, and Nebraskan. But there was no way to know the absolute ages of these episodes nor whether there were still-earlier cycles of ice growth and recession. The problem with land-based records is that each ice advance, like a Zamboni resurfacing the ice between periods in a hockey game, will tend to erode and overprint the record of the previous events. Wisconsin (outside the Driftless Area) was glaciated in all four ice advances, but it is often difficult to recognize the deposits of the earlier three.

In the last years of the nineteenth century, Chamberlin and many others speculated about the causes of the Ice Ages, invoking not only orbital cycles but also volcanism, mountain building, and ocean circulation. In 1896, Swedish chemist Svante Arrhenius made the case that certain trace atmospheric gases, particularly *carbonic acid* ($H_2CO_3$, carbon dioxide combined with water vapor), could be important in governing climate because they are transparent to incoming short-wavelength radiation (light) from the sun but block outgoing long-wavelength energy (heat) reradiated from the Earth's surface.[16] (He even surmised that emissions from coal burning might one day "improve" Sweden's climate). All these ideas would eventually prove to be at least partly correct, but at the time, none could be rigorously tested without higher-resolution information about how climate had changed over time. There were many climate suspects, but it was premature to bring them to trial; the evidence was still too circumstantial.

## ESPRIT DE CORES

Finally, in the 1970s, two new, rich archives of climate data that revolutionized climate science were opened—as if someone doing scholarly work with random volumes in a used bookstore suddenly had access to the Library of Congress. These were (1) deep-sea sediment cores obtained from a new generation of oceanographic research vessels and (2) polar ice, sampled through heroic international drilling operations in Antarctica and Greenland. The deep seafloor and polar ice caps are similar in being sites where slow, continuous accumulation occurs without interruption or disturbance, like dust gradually blanketing furniture in a closed-up room. Today, deep-sea cores from all the world's oceans provide a 160 million-year record of global climate change (extending back long before the Ice Age), encoded as variations in geochemistry and microscopic fossils, at a resolution of thousands of years. Ice cores, in turn, document 700,000 years of atmospheric variations that can be read to the year, at least in young ice. Teasing climate information from seafloor ooze and old snow, however, requires code-breaking—translating the cryptic record of stable isotopes in shell and ice.

Oxygen, like carbon, has two main stable isotopes, and in the same way that light carbon ($^{12}C$) is "preferred" by photosynthesizing organisms over the heavier form ($^{13}C$), light oxygen ($^{16}O$) is more likely to be taken up as water vapor during evaporation than heavier $^{18}O$. This means that at any given time, precipitation, including polar snows, will have less $^{18}O$ and more $^{16}O$ than ocean water, and this sorting effect is further enhanced during cold periods. During ice ages, when a significant fraction of Earth's water is locked up in glaciers and ice caps, the oceans, and the organisms that form their shells from ocean water, will have particularly high ratios of $^{18}O$ to $^{16}O$. Conversely, glacial ice will have particularly low values of this ratio. Ratios of ordinary hydrogen ($^{1}H$) to deuterium ($^{2}H$) vary in a similar way, and so in glacial ice (which is, after all, $H_2O$) there is a second proxy record of climate. Isotope ratios in sea sediments and ice thus provide high-fidelity documentation of both global ice volume and temperature over time.

The ice cores, and the much longer sea-sediment records, reveal that Chamberlin's four ice advances were just the most recent of *30* that occurred over the 2.6 million- year span of the Pleistocene. And the throbbing signal of the Croll-Milankovitch cycles—a strong regular beat, with superimposed flutters—is unmistakable.[17] For the first 1.5 million years of the Pleistocene, the 41,000-year obliquity cycle is especially evident. Then, around 1.2 million years ago, the pulse slows to the calmer 100,000-year rhythm of the eccentricity cycle, like an electrocardiogram readout for a patient who is falling off to sleep. This is called the mid-Pleistocene transition, and its cause is not completely understood. For one thing, of the three orbital variables, eccentricity has the smallest effect on solar radiation received by Earth, but for some reason the 100,000-year cycle became amplified by geologic processes. There are also higher-frequency "harmonics" in the climate records that do not correlate with orbital variations. Recurrent temperature oscillations of about 1500 years—the so-called Dansgaard-Oeschger cycles—seem to be a characteristic internal rhythm corresponding to the timescales of global ocean currents. This means that the planet is not simply a puppet dancing to the imposed rhythms of astronomical cycles but that it takes those rhythms and riffs on them in its own way.

## HEAT OF THE MOMENT

There is an even more important difference between the predicted effects of the combined orbital cycles and the observed records of climate, and it further illustrates Earth's capacity to improvise on themes by Milankovitch. The Milankovitch cycles are all essentially sine waves—symmetrical, palindromic hills and valleys. When superimposed, they create more complex patterns, but overall there is no systematic directionality to them—at a glance, it wouldn't be clear which way time is flowing. The actual climate records from sea sediment and ice, in contrast, have an asymmetric, sawtooth geometry: long periods of cooling when Earth slid slowly into ice ages are punctuated by short and abrupt episodes of warming. That is, in each cycle, a tiny orbital nudge toward warmer conditions was magnified by something in the Earth system into a heat wave, like a thermostat gone haywire. The cause of this amplification is preserved in the ice itself: greenhouse gases, especially carbon dioxide and methane ($CH_4$), or swamp gas.

When snow accumulates, air pockets remain between the crystals (making snow shelters surprisingly warm, because they are well insulated). At the poles, where snow doesn't melt from one season to the next, it compacts as it is buried, and at a depth of about 60 m (200 ft) recrystallizes into ice. In the process, the air pockets shrink, but vestiges remain as bubbles suspended in the ice, like insects caught in amber. While there may be some migration of air between layers in this process, the gas bubbles trapped in polar ice are a direct record of past atmospheric compositions at the resolution of at least decades. These tiny bubbles tell us that over the last 700,000 years, global temperatures have been correlated at the very highest level of statistical significance with the concentrations of the greenhouse gases carbon dioxide and methane.

So how could greenhouse gases take a small increment of Milankovitch warming and magnify it into a meltdown? The answer lies in the many mechanisms of *positive feedback*—self-amplifying processes—in the Earth's climate system. For example, during the long cooling periods of the Pleistocene, areas beyond the margins of the ice sheets would have hosted tundra ecosystems of slow-growing lichens, moss, and small vascular plants, as in Svalbard today. When this vegetation died, the cold temperatures would have inhibited decomposition (largely accomplished by microbial activity, which gets sluggish in the cold), and so organic matter would simply have accumulated over the millennia in thick piles of peat. This fact was scorched into my mind one summer in Svalbard when a colleague and I thought we would clean up an ugly heap of plastic containers and rotting rope that had washed ashore from ships, which often use the ocean as their rubbish bin. We set a fire on the beach and were glad to find that the nasty trash burned well. Then we noticed that a strip of mossy tundra a little further inland was no longer moist and green but instead dry and brown—and seemed to be smoking. In a sickening instant, we realized that our fire had ignited a hidden layer of peat under the beach cobbles. Fortunately, after several frantic minutes of rushing back and forth with cooking kettles between the slow-moving front and the sea, we had doused the smoldering fire.

Fire is dramatic evidence of rapid oxidation; decomposition accomplishes the same thing invisibly, in slow motion. During times in the Pleistocene when Milankovitch cycles caused temperatures to warm even a little, tundra microbes would wake up and get back to work, munching away at the plentiful peat and releasing its long-sequestered carbon as carbon dioxide (or methane where oxygen was scarce). This in turn warmed the planet more, further accelerating the microbial feeding frenzy, which released still more greenhouse gases, and so on, in a classic positive feedback circle.

Other positive feedbacks in the climate system include the albedo, or reflectivity, effect, which had played a powerful role in sending the cooling planet into a “snowball” state at the end of the Proterozoic. But the albedo effect works both ways: once melting begins, the darker color of dirty ice, bare land, or open sea water causes greater absorption of the sun’s heat, leading to more warming, more melting, and the expansion of dark surfaces. This accelerating warming can then amplify carbon-cycle feedbacks even further.

Positive feedback processes can accentuate cooling—for example, windier conditions during glacial times fertilized iron-starved phytoplankton in the oceans with nutritious dust, and when some fraction of their biomass sank to the seafloor without decomposing, atmospheric $CO_2$ was gradually drawn down. But the sawtooth pattern that is so prominent in ice and sediment cores underscores an inescapable asymmetry in Earth's climate system: it takes a lot longer to cool the planet than to warm it up.

## C SICKNESS

For us in the Anthropocene, the urgent questions are, How fast, exactly, did past warming episodes happen, and How high were greenhouse gas levels at those times? The last glacial maximum, when great ice lobes left the Wisconsinan deposits of Chamberlin, occurred 18,000 years ago. At that time, atmospheric carbon dioxide concentrations stood at 180 parts per million (ppm). After that deep-winter state, orbital factors began to favor milder conditions again, and $CO_2$ levels rose, too. Earth then entered a period of steady warming, interrupted by a temporary cold snap between about 12,800 and 11,700 years ago (the *Younger Dryas interval*), which is thought to have been caused by disruption of the Gulf Stream, which conveys warm tropical waters to the North Atlantic, as fresh water from melting ice sheets flooded the North Atlantic. By this point, $CO_2$ levels had risen to about 255 ppm over 6300 years, at an average rate of 0.01 ppm/yr. When the Gulf Stream reestablished itself, it was as if the Earth had made a New Epoch's resolution to adopt an entirely different mode of behavior. In a matter of just decades, around 11,700 years ago (the golden spike for the Holocene), global average temperatures vaulted suddenly to their Holocene values, and Earth left the bungee-jumping days of the Pleistocene behind.

But the transition into the Holocene was a time of massive geographic readjustment. The ice caps shrank, and their meltwaters ponded in vast lakes. Some of these lakes, bounded by fragile ice barriers, drained catastrophically; the peculiar landscape of the Channeled Scablands in eastern Washington State records unimaginably cataclysmic flooding when an ice dam that had impounded a volume of water equivalent to Lake Michigan suddenly failed (sorry, Mr. Lyell). New river systems set to work organizing drainage networks in the lumpy, deglaciated landscapes. In North America, the main tributaries to the Mississippi, the Missouri and Ohio Rivers, mark the edges of the last ice sheet, where the greatest volumes of meltwater had to be processed. As glacial meltwaters found their way back to the oceans, sea level rose hundreds of feet in a few thousand years, flooding coastal areas and changing old river valleys into estuaries. The land bridge that had connected Asia with North America was drowned. Britain became separated from the rest of Europe as the Channel filled. Eventually, though, coastlines stabilized. Weather patterns became regular and predictable. Humans could get down to the business of raising crops and building civilizations.

By about 1800, just before we began to consume ancient carbon fuels in significant quantities, the concentration of atmospheric $CO_2$ had risen to about 280 ppm, only 35 ppm higher than at the start of the Holocene. This suggests that over the course of 11,000 years, Earth's carbon cycle had settled into an equilibrium state in which the carbon exhaled by volcanoes and released from decaying organic matter was about balanced by the carbon inhaled by photosynthesizers and sequestered as limestone. Now and then, small imbalances in the carbon budget threw human societies into periods of famine and conflict.

In the decades after the industrial revolution, we, like overgrown microbes devouring peat, began to gorge on long-stored carbon—first coal, then petroleum and natural gas.

Photosynthesis and limestone precipitation could no longer keep up. An unjust fact about carbon emissions is that while one part of the world—the United States and western Europe—was responsible for a disproportionate share of the twentieth-century output, the whole world suffers the consequences. This is because the mixing time for the troposphere (the lower atmosphere)—the time it takes for turbulent stirring by winds and weather to homogenize the air on a global scale—is relatively short (1 year) compared with the residence time of carbon in the atmosphere (hundreds of years). If the mixing time were long compared with the residence time, then carbon emissions would hover close to the places where they were released—like garbage piling up when trash haulers strike—and might motivate action to curb them. But because our individual emissions are not only invisible but conveniently dispersed around the world, we feel little incentive to curtail them.[18]

By 1960, the level of global atmospheric $CO_2$ had reached 315 ppm—rising as much in 160 years as it had over the previous 11,000—at a rate of 0.22 ppm/yr, more than 20 times the rate in the Late Pleistocene, when Earth began to heat up significantly. In 1990, we breezed passed the 350 ppm mark, which many climatologists consider the upper threshold for maintaining Holocene climate stability—the point at which the juggernaut of positive feedbacks was likely to be set off. By 2000, the $CO_2$ level had reached 370 ppm, rising at a rate of 2 ppm/yr. As I write, we have broken the 400 ppm ceiling, and the rate of increase is still increasing.

In all the yo-yoing of Pleistocene climate, $CO_2$ levels never exceeded 400 ppm. The last time $CO_2$ concentrations were this high was Pliocene time, more than 4 million years ago. And there is certainly no Pleistocene precedent for the rate at which carbon dioxide levels are increasing. The closest analog is a climate crisis 55 million years ago, at the boundary between the two earliest epochs in the early Cenozoic Era: the Paleocene-Eocene Thermal Maximum, known by its less unwieldy acronym, the PETM.

## A DISTANT MIRROR

Like eye-witness reports of an earthquake, sea-sediment cores at dozens of sites around the globe provide vivid accounts of the PETM. The cores all tell of a sharp shock: a sudden 5°–8° spike in temperature, as recorded by oxygen isotope ratios in microfossils; a simultaneous jump in ocean acidity, marked by a crash in the amount of calcitic shell material; and a huge influx of carbon from some biogenic source, as indicated by its unusually high enrichment in $^{12}C$ relative to $^{13}C$.[19] The fossil record speaks of an ocean ecosystem in disarray: many species of plankton suffered serious reductions in numbers, and an extinction in bottom-dwelling microorganisms called benthic foraminifera indicates that even the deep waters of the ocean were affected. These changes in turn triggered a major reorganization of the marine food chain. On land, hotter and more arid conditions forced dramatic migrations of mammal species, while one-fifth of plant species, unable to move fast enough, went extinct. Marine and land-based records of the PETM indicate that it took the oceans and biosphere 200,000 years to achieve a new equilibrium.[20]

The size of the shift in carbon isotope ratios during the PETM allows estimates of the amount of carbon that must have been released; most calculations fall in the range of 2000 to 6000 billion metric tons, or gigatons (Gt), of carbon. (*Note*: Sometimes carbon emissions are reported as Gt of $CO_2$, not just C; in this case, values are greater by a factor of 3.7, reflecting the higher molecular mass of $CO_2$). The 2000–6000 Gt figure is hard to understand until one realizes that total cumulative anthropogenic carbon emissions to date are around 500 Gt, and a quarter of that has been released since the year 2000. With rates of emissions still climbing, we are likely to reach or exceed many of the estimates of the PETM carbon spike by the year 2100.

An important but unresolved question is how so much biogenic carbon could have been released in the PETM, long before humans got into the habit of burning fossil fuels. The two primary candidates are (1) ignition of coal or peat by magmatic activity during the opening of the North Atlantic ocean (akin to the long-burning underground fires that have smoldered for 50 years beneath Centralia, Pennsylvania); and (2) sudden vaporization of a form of methane caged in ice—*clathrate* or *gas hydrate*—from sediments on the seafloor. This frozen methane, produced by microbes happily gobbling up organic matter, is stable under only a limited range of temperature and pressure conditions. If seawater warms, or if a submarine landslide suddenly uncovers a layer rich in gas hydrates, the frozen methane can become unstable and erupt from the seafloor in great oceanic belches. Gas hydrates weren't even known until the 1980s; before that, sediment cores commonly came up with large voids in them, indicating that something had been lost on ascent—the strange ices had vaporized before scientists could even look at the cores. More efficient core recovery finally revealed what had occupied the empty spaces: ice that could be burned. Estimates of the mass of gas hydrate currently stockpiled in marine sediments vary from 1000 to 10,000 Gt. Like tundra peat, these carbon stores could become unstable as climate warms; their sudden volatilization would trigger a nightmarish runaway greenhouse effect.

But the sedimentary record of the PETM, with a resolution no better than a few millennia, does not allow us to distinguish between an essentially instantaneous release of carbon from a belching ocean and a longer-term (1000-year) combustion of coal or peat. This distinction is not of merely academic interest. If the denominator for the rate of carbon output in the PETM is one year, we can still cling to the idea that our emissions are not completely unprecedented. But if the denominator is thousands of years, our Anthropocene carbon spewing is a truly extreme geologic outlier.

## A NEW LEAF

These days, we humans are emitting more than 10 Gt of carbon every year—mainly through fossil fuel burning, but also cement production (which roasts limestone) and deforestation—easily out-gassing the world's volcanoes by a factor of 100. But could we mimic biogeochemical cycles and find ways to take the carbon we emit back out of the atmosphere? There are many possible strategies, ranging from cutting-edge engineering to direct replication of natural processes. So far, the high-tech approaches are too expensive to be feasible and the low-tech ones are too slow; the thing about geologic processes is that they tend to take their own sweet geologic time.

For years, the U.S. coal industry has been pushing the oxymoronic idea of "clean coal," based on the unlikely scenario that carbon capture and sequestration (CCS) systems would be installed in power plants around the country. The technological capability for CCS exists; it involves containing the $CO_2$ emitted from coal combustion, compressing the gas at high pressure, and injecting it into porous rocks deep underground, ideally on or near the site of the power plant (if the local geology is suitable). For power plants near coastlines, some CCS schemes have imagined disposal of $CO_2$ in deep-ocean water, but this would be self-defeating, since ocean acidification is one of the effects of elevated atmospheric $CO_2$ that the sequestration process is trying to mitigate in the first place.

For a time in the early 2000s, it seemed possible that with sufficient economic incentives—such as a carbon tax, or a cap-and-trade carbon emissions market—that CCS technologies might be implemented on a broad scale, but this was quashed by the emergence of "unconventional" natural gas production from shales through horizontal drilling and hydrofracturing, or "fracking." Energy prices fell dramatically, and the lower net $CO_2$ output from combustion of natural gas, compared with coal, drained the momentum from the nascent movement toward CCS. (While it is true that natural gas emits about 50% less $CO_2$ than coal per heat unit produced, the gas industry's claims about natural gas as a low-$CO_2$ fuel are partly negated by "fugitive" methane leaking from poorly sealed wells and badly maintained pipelines.)[21] Gas-fueled power plants could also employ carbon capture systems, but at a steep price: construction costs for new plants would be almost doubled, and the cost of $CO_2$ captured—which sets the lower limit for an effective carbon tax or market value—is estimated at about $70/ton, excluding transport and storage.[22] In the present economic and political climate, CCS seems unlikely to be the solution to the miasma of carbon we have created.

Even if carbon capture technologies were economically viable, they are not necessarily a panacea. Direct $CO_2$ emissions from power plants can be reduced by 80%–90%, but significant amounts of energy are required for the CCS process itself. And if sequestration cannot be done on-site, transport of $CO_2$ creates additional energy demands. Finally, the injection of pressurized $CO_2$ into deep geologic formations is not without challenges. The rocks to be used as a storage "container" must be porous enough to hold large amounts of compressed gas but not permeable enough to allow it to leak out—which is a bit like valuing a friend for his big-hearted gregariousness but then expecting him to keep a juicy secret. And forcing high-pressure fluid into rocks, whether it is $CO_2$ or wastewater from hydrofracturing, can have an unsettling side effect: inducing earthquakes, which, ironically, could compromise the integrity of the carbon dioxide reservoir.

Instead of capturing carbon from power plants, could we mimic photosynthesizers and extract $CO_2$ directly from the air? For at least two decades, a number of academics and private companies have worked on developing "artificial trees" whose "leaves" would bind ambient $CO_2$ in a chemical medium such as a strong base, like lye (sodium hydroxide, NaOH) or a polymer resin. An optimistic advocate for this technology is physicist Klaus Lackner of Arizona State University, who believes it is possible to engineer a "tree" that could capture as much as 1 ton of $CO_2$ per day, about 1000 times more than the average natural tree. At this optimum level of efficiency, it would take 30 million artificial trees to keep up with our current 10 Gt/yr carbon habit, and hundreds of millions more to reverse the effects of a century of carbon emissions—or even get back to the 1990 level of 350 ppm that many climatologists see as a tipping point.

A study by the American Institute of Physics estimates that the cost of direct air capture of $CO_2$, using even the most promising (but still unproven) technologies, would be about $780/ton of $CO_2$, almost 10 times more than carbon capture and sequestration at power plants.[23] Also, direct-capture "forests" would require large land areas, and the carbon they captured would still need to be disposed of either through underground injection or burial in some solid form.

## KNOCKING ON WOOD

All these concepts make old-fashioned photosynthesis seem like an incredible bargain—and we have the technology! So, is planting as many seeds and saplings as possible the solution? As the geologic record shows, the trick to reducing atmospheric $CO_2$ levels is to sequester more carbon through photosynthesis each year than is released by decomposition. (The irony, of course, is that undecomposed organic carbon of the geologic past made the fossil fuels that got us into this predicament today). There is no net change in $CO_2$ levels if carbon fixed by plants in the spring and summer is then released in the fall and winter through their decay. Fast-growing trees with a long lifespan are therefore the darlings of carbon sequestration. While they don't store carbon forever, they can keep it out of circulation for decades or centuries.

But even the simple idea of planting trees to modulate carbon gets complicated in implementation. First, there is obviously a limit to the land area that can be reforested; we do need to grow food (though in the last century, parts of the northern United States such as Wisconsin and New England, which had been clear-cut and farmed in the nineteenth century, are now returning to forest land). Also, one might think that young trees, with vigorous growth rates, would capture more carbon. If this is true, it would make sense to cut down old forests to make space for new plantings. But recent studies have shown, counterintuitively, that many species of trees actually sequester more and more carbon as they age, because their leaf area, girth, and branch volume continue to increase.[24] Letting old trees continue to grow while also planting new ones thus seems the best strategy. Still, trees have a finite lifespan and eventually return their carbon to the atmosphere.

A more active approach to harnessing the power of photosynthesis is known by the functional but cumbersome name "bioenergy with carbon capture and storage" (BECCS). The idea is to use biomass from fast-growing photosynthesizers—plants like switchgrass or "farmed" algae—as a fuel source, and then sequester the carbon emitted in the combustion of this fuel. In theory, this could be a truly carbon-negative process, since at least some carbon extracted by photosynthesis would be withdrawn from the atmosphere for the long term. Small-scale pilot projects have shown promise, but converting plant matter to fuel is itself energy-intensive, and carbon capture at biomass facilities may be even more expensive than for coal or gas.[25]

Over geologic time, much photosynthetic carbon has been sequestered as marine biomass, mostly bacterial, that fell to the seafloor and was buried in low-oxygen sediments (some of which became petroleum, natural gas, or gas hydrates). Perhaps we could emulate this process by stimulating the growth of plankton communities in the oceans, in the hope that some of the carbon they fix will find its way into sediments and be locked away for geologic timescales. The best fertilizer would be iron, which microbes have been starving for since the Great Oxygen Revolution of the Proterozoic.

Intentional manipulation of ocean chemistry, however, raises alarms among marine biologists. Altering the base of the food chain is certain to have negative and unforeseen consequences (we are already doing this unintentionally—but knowingly—by failing to mitigate phosphorus and nitrate runoff from agriculture, which leads to anoxic coastal dead zones). This is why there was scientific outcry in 2007 when entrepreneur Russell George started selling shares in a company called Planktos, which intended to fertilize a Rhode Island–sized area of the Pacific Ocean and sell carbon offsets to environmentally minded consumers. Planktos failed, but George reappeared in 2012 as a consultant to a First Nation in coastal British Columbia, the Haida people, promising to revitalize their anemic salmon fishery with iron fertilization. One hundred tons of iron sulfate were dumped in the waters around the Haida Gwaii (Queen Charlotte) Islands, with inconclusive results, before the UN's International Maritime Organization condemned the act, and the Canadian environmental ministry intervened to stop it. The scientific unease about cavalier alteration of seawater arises partly from the fact that we can't be sure that our current understanding of ocean biogeochemistry will even apply in the near future. We have incomplete knowledge of the global marine microbiome as it exists today and still less of a grasp on how it might evolve as the seas grow warmer and more acidic.[26]

## LIMELIGHT ON LIMESTONE

If accelerating microbial growth in the oceans is off the table, perhaps we could imitate Earth's long-term carbon sequestration scheme: fixing atmospheric carbon dioxide in limestone. Making limestone begins with weathering silicate rocks to free up calcium that can then combine with atmosphere $CO_2$ to form calcium carbonate or calcite. This is the process responsible for the slow drawdown of $CO_2$ that cooled the globe as the Himalaya grew (Ch. 3). In nature, shelled organisms do the work, sopping up carbon at an estimated 0.1 Gt/yr—sufficient over geologic time to have locked into solid rock 99.9% of all the carbon dioxide emitted by volcanoes, but 100 times too slow to keep up with our current annual emissions. And unfortunately, making shells will become an even harder task as ocean acidity increases, causing the already slow natural rate of limestone formation to decrease in the coming centuries.

It might be possible, however, to form artificial "limestone" by deliberately facilitating the silicate weathering reactions that draw $CO_2$ out of the air. An igneous rock type called peridotite, rich in the mineral olivine (whose gem form is peridot) will react with carbon dioxide to form a magnesium-rich carbonate mineral (magnesite) similar to calcite, as follows:

$$Mg_2SiO_4 + 2CO_2 \rightarrow 2MgCO_3 + SiO_2$$
$$\text{Olivine} + \text{Carbon dioxide} \rightarrow \text{Magnesite} + \text{Quartz}$$

The catch is that although peridotite is very abundant in the Earth—it makes up most of the Earth's upper mantle—it is quite rare on Earth's surface. But there are places, including Newfoundland, Oman, Cyprus, and Northern California, where subduction went wrong and slabs of mantle rock were thrust up onto the edges of continents. At these locations, peridotite could be perforated with drill holes into which captured $CO_2$ could be pumped. One study suggests that the Oman peridotites alone could sequester 1 Gt of carbon per year (one-tenth of our annual output).[27] The carbonation reaction is sluggish at low temperatures, but it is also exothermic, so once it begins, it is self-accelerating. The main problem, of course, is getting the gas to the rocks. Carbon dioxide must either be captured and transported to the rare places where mantle rocks lie at the surface, or peridotite must be mined in large volumes and spread over vast areas of Earth's surface, where it could react passively with the atmosphere.

## AIR RAIDS

Given all the difficulties with getting rid of carbon dioxide, it is no wonder that the idea of cooling the planet by shooting sulfate aerosols into the stratosphere—inspired by the 1991 eruption of Mount Pinatubo—is so seductive. "Solar radiation management" is relatively cheap (billions of dollars a year) and could probably be started right away using rockets, airships, or high-altitude jets. But it would also be a Faustian bargain. Once begun, a sulfate injection scheme would require a decades-to-century-scale commitment, since in the absence of serious $CO_2$ reductions, it would mask but not reverse greenhouse warming (ocean acidification from rising $CO_2$ levels would also continue unabated—and undermine carbonate precipitation, Earth's slow but effective long-term carbon sequestration system). There is also the moral hazard that suppressing the symptom would reduce the political will to cure the underlying disease. Stopping injections after a period of a few years would lead to ferocious "catch-up" warming that could devastate the biosphere and lead to extreme alteration of weather patterns.

Adding a Pinatubo-equivalent mass (about 17 megatons) of sulfur dioxide to the stratosphere every few years for 50 or 100 years would fundamentally change biogeochemical cycles in ways we can only partly anticipate. And, like a drug addict needing larger and larger doses to get the same high, the amount of sulfate required to attain the same level of cooling would actually increase over the years. This is because both the residence time and reflectivity of the sulfate droplets would steadily decrease as a result of their tendency to glom together and grow larger; bigger particles fall out of the atmosphere faster, and they have smaller surface area relative to their volume than small ones, which reduces their efficiency as solar energy reflectors.

Atmospheric chemists do know that large volumes of stratospheric sulfate would damage Earth's radiation-shielding ozone layer, which has been slowly recovering since 1989, when the Montreal Protocol first limited the use of chlorofluorocarbons. Also, the environmental impact of the sulfate delivery system would itself be considerable: if jet fighters were used, millions of flights would be required each year.[28] And for each mission to launch sulfate into the stratosphere 10 km (6 mi) up, there is a possibility that the payload could fail to reach the target altitude, unleashing a localized downpour of acid rain.

A sulfate shroud would alter the wavelengths and intensity of light that falls on photosynthesizing plankton and plants, with unknown effects on natural food webs, forests, and agricultural crops. A particularly cruel irony is that aerosols would reduce the efficiency of solar power generation, especially large-scale solar arrays that use mirrors and lenses to concentrate sunlight, thereby undercutting a technology that could help wean us from the fossil fuels that are the root of the climate problem.[29] Because sulfate aerosols have no effect in the dark, when there is no light to reflect, they would reduce day/night, summer/winter, and tropical/polar temperature differences. This would likely cause dramatic shifts in global weather patterns, which are driven by temperature contrasts and gradients. The potential effects on the many complex temperature-driven ocean-atmosphere interactions like the interannual El Niño cycle and the monthly to bimonthly *Madden-Julian oscillations*, which govern weather around the Pacific basin, are unclear. Multiple climate models suggest that areas affected by the annual Asian monsoon could see sharp reductions in precipitation, although there are large uncertainties in these simulations.[30] What recourse would there be for regions adversely affected by atmospheric manipulation? Given the state of world governance, it is hard to imagine that this intergenerational global geochemical experiment could be smoothly administered and promote harmony among nations. And did anyone mention that the sky would always be white, not blue?

It is telling that the most vocal advocates for stratospheric sulfate injection are either economists, accustomed to viewing the natural world as a system of commodities whose "real" value is in dollars, or physicists, who treat it as an easily understood laboratory model. Often, the argument is made that our unintentional atmospheric modification from greenhouse gas emissions has now reached the point where there is "no choice" but to perform intentional "management" of climate.[31] Most geoscientists, knowing the long and complex story of the atmosphere, biosphere, and climate—the hellish extinctions and feverish ice ages, fragile food chains, and powerful feedback mechanisms—think the idea humans can "manage" the planet is delusional and dangerous. What on Earth makes us think we can control nature on a global scale, when we haven't even learned to control ourselves?

## BACK TO NATURE

The carbon conundrum is not the only environmental challenge of our time, but it underscores a more general, obstinate fact: that there is an immense asymmetry in the time it takes to consume, alter, or destroy natural phenomena compared with the time required to replace, restore, or repair them. This is the hard truth I first glimpsed in the shards of a tourmaline crystal, and it is the central challenge of the Anthropocene.

This brave new epoch is not the time when we took charge of things; it is just the point at which our insouciant and ravenous ways starting changing Earth's Holocene habits. It is also not the "end of nature" but, instead, the end of the illusion that we are outside nature. Dazzled by our own creations, we have forgotten that we are wholly embedded in a much older, more powerful world whose constancy we take for granted. As a species, we are much less flexible than we would like to believe, vulnerable to economic loss and prone to social unrest when nature—in the guise of Katrina, Sandy, or Harvey, among others—diverges just a little from what we expect. Averse to the even smallest changes, we have now set the stage for environmental deviations that will be larger and less predictable than any we have faced before. The great irony of the Anthropocene is that our outsized effects on the planet have in fact put Nature firmly back in charge, with a still-unpublished set of rules we will simply have to guess at. The fossil record of previous planetary upheavals makes it clear that there may be a long period of biogeochemical capriciousness before a new, stable regime emerges.

CHAPTER 6

# TIMEFULNESS

## UTOPIAN AND SCIENTIFIC

The distinction between the past, present and future is only a stubbornly persistent illusion.

—ALBERT EINSTEIN, IN A LETTER TO MICHELE BESSO, 1955

### LEVIATHAN

For a few weeks each February, small towns pop up like Brigadoons on the ice of Lake Winnebago, the largest inland water body in Wisconsin. Winnebago is a vestige of the much larger glacial Lake Oshkosh, which formed from ponded meltwater late in the Ice Age and left behind heavy clay sediment that is the bane of gardeners in our area. Lake Winnebago is shallow, and often alarmingly green in the summer as a result of runoff from lawns and farms, but it still supports a healthy population of lake sturgeon. Each year, before they head into upstream tributaries to spawn, the sturgeon in Lake Winnebago congregate in a few areas, and the temporary towns start to appear on the ice, mirroring the fish communities below.

Sturgeon are large fish—the record setter for this area was 240 pounds (bigger, as the local paper pointed out, than a popular Packers linebacker[1]). Their lifespan is longer than that of humans, and their lineage has been around since the Early Cretaceous. They are caught not with delicate hooks and lines dropped through narrow auger holes but with trident-like spears plunged into large rectangular openings sawn in the ice. If spearing sounds brutal, it is at least a fair match between humans and fish. Spearers wait for hours or days in dark shanties illuminated only by the otherworldly glow of secondhand sunlight that shines through the ice and reflects off the bottom of the lake. If a sturgeon happens to swim by, it is a feat of athleticism to plunge the spear with sufficient force at the precisely right moment to strike it, and then to wrestle it out of the frigid water. Some people have sat in sturgeon shacks for 30 seasons without getting a single fish. Some fish have been swimming in the lake for more than a century without being caught.

As early as the 1910s, there was concern about the declining sturgeon population in Lake Winnebago and connected waters. Both the flesh and the roe of sturgeon fetched high prices, and year after year, commercial fishing operations caught as many fish as possible. In the winter of 1953, when almost 3000 fish were taken, the public awoke to the possibility that the sturgeon could soon be harvested to extinction. Sturgeon spearers and the Wisconsin Department of Natural Resources began to work together to monitor the population and set catch limits.[2] During the spring spawning season, citizen volunteers (the "Sturgeon Patrol") stand guard along the tributary rivers where the females come halfway out of the water to lay their eggs on shallowly submerged rocks, and males follow to fertilize them. DNR biologists keep a close watch on the winter sturgeon harvest. As soon as the quota for a given year is reached, the season ends, sometimes just hours after it opened, and spearers, knowing this protects the sturgeon stock for the future, respect the system. Weigh stations are set up on shore at the spots where the ice roads to the shanty towns begin. Each fish is sexed and weighed, and its age is estimated by cutting a slice of its dorsal fin, which has growth bands like tree rings. *That one's older than great-grandma! This one hatched when Coolidge was President!* The weigh station is itself an ephemeral village, where people of all ages gather to see the giant fish pulled from this parallel, primeval world that exists so close by but can be glimpsed for only a few weeks each winter.

## IN SEARCH OF LOST TIME

French philosopher Bruno Latour has argued that a defining characteristic of modern society is "a peculiar propensity for understanding time that passes as if it were abolishing the past behind it.[3]" We think that our worldview represents an "epistemic rupture so radical that nothing of the past survives" in it and that our technologies lift us above the oppression of natural history that for so long defined the human experience. As permanent exiles from the past, we have mixed emotions about it. We allow ourselves moments of nostalgia but scold people for "living in the past." The prevailing consensus is that the past must in fact be abolished to make way for better things (Do you remember those old flip phones?). We caution each other about becoming Luddites, slipping backward, returning to the dark ages.

But stranded on the island of Now, we are lonely. When I see people crowded together in the cold each year to see big, old, ugly fish being weighed, I sense a very unmodern yearning to connect with the past. And I suspect that our self-imposed exile from it is the source of many ills: environmental malefactions and existential malaise are arguably both rooted in a distorted sense of humanity's place in the history of the natural world. People would treat each other, and the planet, better if we embraced our shared past and common destiny, seeing ourselves more as lucky inheritors and eventual bequeathers rather than permanent residents of the Earth estate. In short, we need a new relationship with time.

Our modern conviction that time is a one-way vector and the past is irretrievably lost itself represents a break with the past. Earlier societies and cultures were permeated with the presence of ancestral spirits and the practice of ancient rituals that knitted the living, the dead, and the not-yet-born together into a unified temporal fabric, blurring the concepts of past, present, and future. The Buddhist concept of *sati* is typically translated as "mindfulness," or being attentive only to the Now, but it actually means something closer to "memory of the present"—that is, awareness of this moment from a vantage point outside it.[4] The Ghanaian idea of *sankofa*, usually symbolized by a backward-looking bird, is a reminder to move forward but also keep the past in view. In Norse mythology, Ygdrassil, the World Tree that holds up the cosmos, is maintained by three women, the mysterious Norns, called Urər, Verəandi, and Skuld. Sometimes interpreted as Past, Present, and Future, their names literally mean "Fate, Becoming, and Necessity," suggesting a strange, circular conception of time in which the future is embedded in the past.[5] Each day, the Norns nourish the tree from the sacred Well, which holds ancient waters, and recite the Orlog, the eternal laws that have always governed the world. Both acts embody the Norse idea of *wyrd*, or the power of the past upon the present.[6]

In many ways, geology is about understanding "wyrd"—the ways that the secret stories of the past hold up the world, envelop us in the present, and set our path into the future. The past is not lost; in fact, it is palpably present in rocks, landscapes, groundwater, glaciers, and ecosystems. Just as one's experience of a great city is enriched by an understanding of the historical context of its architecture, there is deep satisfaction in recognizing the distinctive "styles" of past geologic periods. And we, too, dwell in geologic time.

I often feel I live not just in Wisconsin but in many Wisconsins. Even when I try not to, I can't help but sense the lingering influence of the many natural and human histories embedded in this landscape: the forests still recovering from nineteenth-century clear-cutting; the rivers that governed ancient trade routes, themselves shaped by moraines shoved up by the great ice sheets; the golden sandstones marking the shores of the Paleozoic seas; contorted gneisses that are the surviving roots of Proterozoic mountains. The Ordovician is not a dim abstraction; I was there with students just the other day! For geologists, every outcrop is a portal to an earlier world. I am so accustomed to this "polytemporal" way of thinking that I'm caught by surprise when I'm reminded that it's not the norm.

Wisconsin is a water-rich state, bounded by two Great Lakes, dotted by thousands of smaller ones, veined by rivers, and graced with reliable aquifers that are refreshed each year by rain and snow. But the growth of urban areas and corporate farms has led to groundwater crises in some parts of the state. Until recently, state law limited the installation of high-capacity wells to areas where natural replenishment rates could keep pace with withdrawals. Depending on the nature of the local rocks or glacial sediments, natural groundwater flow rates can range from feet per day to feet per year, and depending on the depth of a well, the groundwater it taps may have been there for years, decades, or centuries. So knowing both the geologic backstory and the human history of groundwater use in an area is critical to maintaining aquifers. But the state's business-minded attorney general has ruled that the Department of Natural Resources does not have the authority to consider the compounding effects of wells in any given area, arguing that it is "unfair" for the DNR to issue a well permit for one industrial dairy operation and then deny a permit to another.[7] In so doing, the attorney general decreed both past and future irrelevant. Only the present matters.

An irony of our technological advancement is that it has created a society that is in many ways scientifically more naïve than the preindustrial world, in which no citizen who learned physics through backbreaking work and understood climate through subsistence agriculture would have assumed that he or she was exempt from the laws of nature. The "modern" kind of magical thinking is characterized by the belief that repeating falsehoods like incantations can transform them into scientific truth. It is also yoked to a quasi-mystical faith in the free market, which, according to the prophets, will somehow allow us to live beyond our means indefinitely.

The problem, in essence, is that rates of technological progress far outstrip the rate at which human wisdom matures (in the same way that environmental changes outpace evolutionary adaptation in mass extinction events). Critic and author Leon Wieseltier contends that "every technology is used before it is completely understood. There is always a lag between an innovation and the apprehension of its consequences."[8] The rapid obsolescence of digital technologies and the cultural flotsam they deliver corrodes our respect for what lasts ("That was so five minutes ago"). And just as reliance on GPS navigation systems causes our capacity for spatial visualization to atrophy, the frictionless, atemporal instantaneity of digital communications weakens our grasp on the structure of time. Our "modern" idea that only Now is real is arguably delusional, while the medieval concept of "wyrd" seems positively enlightened. And our blindness to the presence of the past in fact imperils our future.

## LIKE THERE'S NO TOMORROW

It will not be easy to break the habit of thinking about Now as an island separated by wide straits from the rest of time. We like our Now—the way the insistent chimes of our digital devices keep us from dwelling too much on the past or planning too carefully for the future. A lifetime's exposure to advertising has allowed the corporate promise of eternal youth to burrow deep into our brains, impelling us to buy the next novel thing, to maintain the illusion that we are exempt from the passage of time and that this Now will never end. The highest-compensated workers in our culture are hedge fund managers, rewarded for writing algorithms that make decisions on timescales of seconds—now and now and now.

These days, a Google search for "Seventh Generation" returns links to the official website and social media accounts of the cleaning products company by that name (now owned by Unilever, a multinational corporation). But the Seventh Generation idea, articulated more than 300 years ago in the Iroquois *Gayanashagowa* (the "Great Binding Law" or "Great Law of Peace"[9]), remains as radical and visionary as ever: that leaders should take actions only after contemplating their likely effects on "the unborn of the future Nation . . . whose faces are yet beneath the surface of the ground." Seven generations, perhaps a century and a half, is longer than a single lifetime but not beyond human experience. It is the span from one's great-grandparents to one's great-grandchildren. From the standpoint of the Seven Generations principle, our current society is a kleptocracy stealing from the future. What would it take for this old idea to be adopted in a modern world that does not even acknowledge time?

What do we owe the future? After all, as the bumper-sticker bon mot goes, "What have future generations ever done for us?" The philosopher Samuel Scheffler posits that they actually do a lot. He points out that if we knew that the human race would die out soon after our own death, our experience as humans would be radically different: "The knowledge that we and everyone we know and love will someday die does not cause most of us to lose confidence in the value of our daily activities. But the knowledge that no new people would come into existence would make many of those things seem pointless."[10] Inspired by the plot of P. D. James's dystopian novel *Children of Men*, Scheffler suggests that our capacity to live full lives depends on the belief that we occupy "a place in an ongoing human history, in a temporally extended chain of lives and generations."

So as thanks for keeping us sane, how can we compensate future generations? From a purely economic standpoint, we should invest in preventing future environmental problems as long as the future benefits are greater than the present costs—and every economic study of the expected effects of climate change indicates that any investment now will repay itself manyfold. The real problem is shifting the timeframe for economic decision-making from fiscal quarters to decades or longer. In a provocative paper published in *Nature*, "Cooperating with the Future," a group of economists and evolutionary biologists developed a model, in the form of a game, to identify economic incentives or governance strategies that might encourage intergenerational decisions about resource use.[11] In the game, they found that a resource is almost always depleted within one generation if decisions are made at the individual level, usually by one or two "rogue" players who extract more than what the others consider a fair or reasonable share. This is of course, the classic *tragedy of the commons*—the despoiling of a collective resource (like a pasture) that could be maintained indefinitely through collective restraint if not for selfish behavior by a minority of bad actors (shepherds who graze too many sheep).[12]

But the "Intergenerational Goods Game" found that if each generation was allowed to vote on the amount of the resource that would be extracted in their lifetime, and each player was then allotted their share of the median amount suggested in the vote, at least some fraction of the resource was passed down through multiple generations. Voting enables fair-share takers, who are usually in the majority, to restrain the bad actors. It also helps convince those who might be tempted, in an unregulated system, to violate the commons that it is in their own best interest to observe the collectively determined limit. The system works, however, only if the voting is binding. It didn't require game theory and statistical analysis for the Iroquois Confederacy to figure this out.

## BIG TIME

Our problem is that we lack both the appetite and political-economic infrastructure for intergenerational action. The habit of blinkered thinking is hard to break, but a group of time-transcending art projects may serve as inspirations. Photographer Rachel Sussman[13] traveled around the world to take formal portraits of living organisms older than 2000 years (the real Millennials): a brain coral that has been alive since the time of Plato; baobabs and bristlecone pines that were seedlings when Stonehenge was built; Australian stromatolites doing what they have done since the Proterozoic; Siberian soil bacteria that slumbered for 700,000 years, through six ice advances, now reawakened by Anthropocene warming. These Old Ones open our eyes to alternative relationships with time. They help us, vicariously, to see beyond the horizon of our own mortal limits.

The work of Japanese conceptual artist On Kawara explored *chronos*—the raw experience of time, stripped of narrative.[14] Between 1966 and 2013, he created a series of thousands of paintings collectively called *Today*, which consist only of the date painted in white on a uniformly colored background. From 1970 to 2000 he sent hundreds of telegrams to art dealers and friends all bearing the message "I Am Still Alive" (in this case, the project outlived the medium). In the notes to accompany his exhibitions, he would give his age as the number of days he had lived up to the opening of the show. His 20-volume piece *One Million Years* is a list of dates from 998,031 BC to AD 1,001,997 (a million years before and after 1997). Much of the first half of the work overlaps with the (arguably more interesting) ice-core record from Antarctica. Public readings of *One Million Years* are still being made and recorded in an ongoing project; at the fluent rate of 100 numbers per minute, it would take seven 24-hour days to count to a million.

Katie Paterson's "Future Library" project in Oslo pairs humans and trees as artistic collaborators in a meditation on *kairos*—time imbued with meaning. A committee, whose current members will eventually die and be replaced, is charged with selecting one author to submit a short story each year for the next century (Margaret Atwood was the first). The manuscripts will be stored, unread, in Oslo's Deichmanske Library. Meanwhile, in a specially planted forest north of the city, fir trees are growing. In 2114, when they are 100 years old, they will be harvested and used to make paper on which the stories will be printed as an anthology. The project is underwritten by a trust that will allow its continuation after the people who initiated it are gone.

An organ work by experimental composer John Cage, "ORGAN2/ASLSP (As SLow aS Possible)" is being performed in a 639-year concert in the fourteenth-century cathedral in Halberstadt, Germany.[15] Since the piece began in September 2001 (on Cage's 89th birthday) there have been only a dozen chord changes. Each chord is sustained over periods of months to years by applying weights to the pedals. As in the case of the Future Library, this centuries-long concert will require the cooperation of people across multiple generations.

Inventor Daniel Hillis designed a "10,000 Year Clock" that is being built inside a mountain in western Texas by The Long Now Foundation.[16] Powered by stainless steel bellows that expand and contract as the outside air temperature varies, the clock will have a 10-ft corrosion-resistant titanium pendulum and a sapphire window through which it will detect the Sun's position in the sky and periodically correct itself. Hillis points out that designing an object to last as long as the span of human history necessarily makes one think very differently about time. For example, over 10,000 years, ignoring leap seconds would cause the clock to be off by 30 days. In that time, Earth will be at the opposite extreme of its precession cycle, with the Northern Hemisphere tilted toward the Sun on what is now the winter solstice. Internal environmental changes over that timescale must also be considered. If climate change accelerates and ice caps melt, Earth's orbit will be subtly affected by the transfer of mass from the poles to the oceans.[17]

It may be tempting to dismiss these projects as gimmicks or follies, but their purpose is to reframe the way we think about ourselves in time. They may even provide templates for how we might design infrastructures for intergenerational governance. At present, hardly any public or even private entities are configured in a way that allows planning on timescales longer than an election cycle or a few fiscal years. The increasing concentration of global wealth in the hands of a tiny minority means that for most of the world, short-term survival always takes priority over preparing for the future. Private philanthropic foundations built from the fortunes of the super rich do have the luxury of thinking on generational timescales and can undertake humanitarian projects that may require decades of sustained effort. Their work is undeniably laudable but it is also deeply undemocratic; it means that a small number of extremely wealthy people are the only ones in charge of the future. And some of them have delusional ideas about it.

Growing numbers of the super rich are investing in lavish "climate bunkers"—the twenty-first century version of fallout shelters—where, in the event of climate catastrophe, they can live out their days in comfort while the rest of humanity deals with scorching heat, encroaching seas, and failing crops.[18] Many of these people are Silicon Valley billionaires whose high-tech companies would seem to be predicated on optimism for the future. Instead, their plan seems to be to sell that illusion to the masses while quietly preparing themselves for apocalypse. Moreover, among the super wealthy there are also starry-eyed futurists who confidently assert that terraforming Mars is a real possibility when the time comes to abandon this planet—and is even the natural and inevitable extension of the human quest for new frontiers. This thinking reveals profound temporal dysmorphia—a deranged understanding of time: not only complete ignorance of the long coevolution of Earth and life but also willful denial of our own history as a species. When have we humans ever been able to execute, over many centuries, a constructive international project (i.e., something other than the devastation of aboriginal civilizations) that required immense expenditures without immediate payback? And how can we imagine that we might prosper on a planetary body with which we have no evolutionary connection? We haven't even learned to take care of each other on this old, friendly, hospitable planet.

At the other end of the economic spectrum, a different model for long-view leadership comes from Native American tribes that have managed to persist—despite centuries of genocide, treaty violations, and grinding poverty—through what cultural theorist Gerald Vizenor calls "survivance." To Vizenor, an enrolled member of Minnesota's White Earth Ojibwe, "Survivance is the continuance of stories . . . the heritable right of succession" rooted in a deep ancestral attachment to land on small reservations—finite worlds.[19] It values endurance over conquest; restraint over consumption; continuity over novelty. It is stubborn, ironic, and self-deprecating, with a clear-eyed view of both the benevolence and capriciousness of Nature and the best and worst of human nature.

In recent years, many Native American tribes have emerged as leaders in environmental stewardship, collecting long-term data sets, organizing grassroots protests, and launching legal challenges to mines and pipelines that threaten public waters. Tribes in Wisconsin, Minnesota, and Michigan pool their resources in the Great Lakes Indian Fish and Wildlife Commission (GLIFWC, or "Glifwick"), which works as a hub that also helps nonnative environmental organizations coordinate legal actions, public education, and conservation initiatives.[20] When the governor declared: "Wisconsin is open for business," and the state legislature gutted four decades of science-based environmental laws in a matter of months, GLIFWC spoke out for the Public Trust Doctrine, which obligates the government to protect lakes and rivers for the collective good. There is a profound, tragic irony in that after so many years of maltreatment by the U.S. government, these tribes are in many ways the truest patriots, committed to saving America from itself.

## FUTURE TENSE

When we peer into the geologic future, a paradox emerges: to some extent, we can see what lies in the far distance more clearly than what is in the foreground. The Sun, as a G-type star, is about halfway through its life expectancy, and in 5 billion years or so will enter its red giant phase, engulfing the Earth and other inner planets. Three billion years before that, however, the Sun's increasing luminosity will lead to an extreme greenhouse effect from the vaporization of Earth's oceans. Once the planet's water is lost to space, the carbon-silicate weathering system that has acted to sequester volcanic $CO_2$ over geologic time will shut down, creating an even more intense greenhouse state that will likely make surface conditions intolerable for all life about 2 billion years from now.[21] The Earth's tectonic system, whose character is intimately wrapped up with the presence of water, will also be profoundly changed. Seawater carried into the mantle with subducting slabs will allow arc volcanism to continue for a few hundred million years after surface water disappears. But without the cooling effect of ocean water, ocean crust will stay hotter and more buoyant longer, inhibiting subduction and altering the pace of tectonics.

For at least the next billion years or so, plate tectonics will continue to shuttle continents to new positions around the globe. The Atlantic Ocean will begin to close, and in about 250 million years, the Americas will be reunited with Europe and Asia in a new supercontinent that has already been named "Pangaea Ultima" by geophysicist Christopher Scotese.[22] Meanwhile, rivers will have erased the Himalaya, Alps, and Rockies.

In about 80,000 years the Earth will reach the point in its Milankovitch eccentricity cycle at which another ice age could happen, but this will depend on greenhouse gas concentrations, ocean circulation, the state of the biosphere, and many other variables. The next thousand years—the same amount of time that separates us from the Viking age—are even harder to bring into focus. If human carbon emissions have not been sharply curbed, and powerful positive feedbacks in the climate system are activated, the Earth could experience a replay of the Paleocene-Eocene Thermal Maximum. Sea level would rise tens of feet, inundating many of the world's most populous cities. Altered weather patterns—more ferocious storms, longer and deeper droughts—would stress world food production. Increasing proportions of government budgets would have to be channeled into crisis management. The balance of geopolitical power would shift depending on how well nations were faring in the new climate regime.

But none of this is foreordained. We have the power to write a different saga for the coming millennium. Rather than lapse into existential despair that we won't be here in a billion years, let us reclaim at least the next few centuries.

## CHRONOTOPIA

It is empowering (or at least therapeutic) in these dark times to imagine what a time-literate society might look like. In his last public interview, Kurt Vonnegut said: "I'll tell you . . . one thing that no cabinet has ever had is a Secretary of the Future, and there are no plans at all for my grandchildren and my great-grandchildren."[23] Let us adopt Vonnegut's suggestion as our first proposal: a representative for the yet-to-be born to serve among the top advisors to the president. The Department of the Future would set in motion a realignment of priorities in all aspects of society. Resource conservation would again become a core value and patriotic virtue. Tax incentives and subsidies would be rebalanced to reward long-term stewardship over short-term exploitation. Putting a price on carbon might help us get a grip on our fossil fuel addiction, sober up, and let us prepare for natural disasters that will happen without our assistance—like the hundreds of large earthquakes that will happen in the next century—rather than expending resources on self-created climate catastrophes.

Poverty and class-based disparity of opportunity would be recognized as problems with deep historical roots that cannot be solved without sustained commitment over commensurate timescales into the future. Public school teachers and others whose work represents an investment in the future would be paid well and held in high esteem. Geology would be fully integrated into science curricula, perhaps serving as a capstone course in which students would apply concepts of physics, chemistry, and biology to the immensely complex Earth system. With a solid understanding of how the planet works, students would go on to become educated voters who would hold public officials accountable for wise governance of water, land, and air. Legislators, governors, and mayors who embrace the Seventh Generation principle would point proudly to what they are working toward and be reelected by grateful constituents.

More generally, schools would help develop children's knowledge of and appetite for history and natural history, instilling in them a deep instinct for their place in Time and a keen curiosity to understand more. The dramatic narratives of the geologic past are perfectly suited to the human appetite for storytelling. A noteworthy psychological study suggests that resistance to the concept of evolution is rooted more in existential dread than religious doctrine, and that it declines as people become more familiar with stories from the natural world.[24] A series of controlled experiments showed that reminders of mortality make many people—across a wide spectrum of religious beliefs—more likely to rate tenets of creationist "intelligent design" favorably, presumably as a source of reassurance in the face of psychological threat. But the investigators also found that the same people, after reading short nontechnical pieces on natural history, were less susceptible to antievolution assertions and seemed to find similar comfort in scientific narratives. As Darwin wrote so lyrically in the closing lines of *On the Origin of Species*:

> There is grandeur in this view of life, with its several powers, having been originally breathed into a few forms or into one; and that, whilst this planet has gone cycling on according to the fixed law of gravity, from so simple a beginning endless forms most beautiful and most wonderful have been, and are being, evolved.

That grandeur has always included us; we have simply tormented ourselves with the idea that we are outside the garden.

In 1973, geneticist Theodosius Dobzhansky, exasperated with "scientific creationists" who were attempting to influence the content of biology textbooks, wrote a classic essay called "Nothing in Biology Makes Sense Except in the Light of Evolution."[25] That title has become a useful polestar for generations of natural science students. In the 1990s, popular writers like Richard Dawkins and Susan Blackmore expanded the scope of evolutionary thinking with the idea of "Universal Darwinism," introducing the concept of the *meme* as the cultural equivalent of the *gene* (though now the term has itself [d]evolved to mean cat videos and images with all-uppercase captions). Theoretical physicist Lee Smolin goes even further and suggests that evolution is literally Universal: he posits that natural selection acting on a population of precursor universes may be the explanation for the improbably well-tuned values of fundamental physical parameters that allow the Universe to exist stably over billions of years. Physical "constants," like the adaptive traits of organisms, may therefore have evolved over time.[26] While Smolin's ideas are not universally (so to speak) accepted in the cosmological community, it is fascinating to see Darwinian thinking entering realms that once exempted themselves from temporality.

While scientists see that everything in nature is connected by the continuous thread of evolution, successive generations of humans are increasingly cut off from each other by the technologies they use—and the cultural memes they trade. We have few institutions in which people at all stages of life can gather and experience a unified sense of human community, what Sigmund Freud called an "oceanic feeling"[27] and philosopher and religious theorist Émile Durkheim termed "collective effervescence."[28] We need spaces where, from an early age, children see that they are on an ancient, sacred path that stretches across time, that the richness of life comes from the universal process of unfolding (e-volution), and that growing up and growing old are to be celebrated, not feared. Religious organizations have traditionally filled this role, but we need to be deliberate about finding new venues—choirs, community gardens, cooking schools, oral history projects, bird-watching groups, sturgeon fishing clubs—that can serve as "intergenerational commons."

In my own career, I've forged deep friendships with people generations older and younger than myself, from many countries and cultures, over our common passion for geology. We've scratched our heads over strange rocks, marveled at stunning vistas, linked arms to ford rushing streams, shared dubious concoctions cooked on tiny camp stoves. It's interesting to note that while prominent scientists in other fields tend to make their most revolutionary contributions in their 20s, geologists mature more gradually, often doing their most important work late in their careers, after a lifetime spent in the company of rocks.

The evolution of geology as a discipline has been similar. Simplistic Victorian ideas about the planet—the dogma of strict uniformitarianism, the belief in fixed continents, the denial of mass extinctions—have given way to a subtler, humbler understanding of an Earth that has many moods and miens, and still harbors deep secrets. For me, geology points to a middle way between the sins of narcissistic pride in our importance and existential despair at our insignificance. It affirms a teaching attributed to the eighteenth-century Polish Rabbi Simcha Bunim that we should all carry two slips of paper in our pockets: one that says "I am ashes and dust," and one that reads "The world was made for me."

The Earth itself, with its immensely deep history, is a communal heritage and universal mentor that may help us find a set of shared values. Studying its past may cause us to reconceive ourselves as fellow citizens of a profoundly mysterious planet that we urgently need to know better. And with leadership from the Secretary of the Future, we can learn to adjust our pace to the tempos of the Earth, repeal the Anthropocene, and reinstate uniformitarianism.

## THE FULLNESS OF TIMEFULNESS

Like many people who experienced childhood—or parenthood—in the past half-century, I love Maurice Sendak's classic book *Where the Wild Things Are*, an allegory about the power of imagination to transport us to other worlds, to transcend time, and to save us from our worst selves. I think of Max's voyage when I teach "History of Earth and Life"—a course with the audacious goal of telling the 4.5 billion-year story of the planet in one academic term (at a clip of about 400 million years a week). It feels as if I have embarked on a long trip with my students. We tour alien landscapes, watch continents move, witness biogeochemical revolutions, asteroid impacts, ice ages, and extinctions, marvel at the profusion of Wild Things, and finally begin to glimpse features that look a bit like home, like Max's room gradually shedding its vines and revealing his bed and table.

We arrive at the present (if I've paced myself properly), with a feeling of exhausted exhilaration, mindful that this world contains so many earlier ones, all still with us in some way—in the rocks beneath our feet, in the air we breathe, in every cell of our body. Geology is in fact the closest we may get to time travel. From our vantage point in the present, we can replay the past at any speed and envision possible futures. This geologic habit of mind—the practice of timefulness—is a fusion of *wyrd* and *sankofa* (sensing the presence of the past), *sati* (holding a memory of the present), and Seventh Generation thinking (a kind of nostalgia for the future). It is something like the way parents see their growing children, poignantly remembering them at earlier stages while holding aspirational visions for who they will become.

If widely adopted, an attitude of timefulness could transform our relationships with nature, our fellow humans, and ourselves. Recognizing that our personal and cultural stories have always been embedded in larger, longer—and still elapsing—Earth stories might save us from environmental hubris. We might learn to place less value on novelty and disruption, and develop respect for durability and resilience. Understanding how historical happenstance is written into each of our personal lives might cause us to treat each other with more empathy. And a timeful, polytemporal worldview might even make us less neurotic about the fact of our own mortality by shifting our focus from the finite length of our life to the rich anthology of experiences that a lifetime represents. While other senses may be dulled with age, the sense of time—which can be developed only by experiencing it—is heightened. Understanding how things have come to be the way they are, what has perished, and what has persisted makes it easier to recognize the difference between the ephemeral and the eternal. Growing old requires one to shed the illusion that there is only one version of the world.

As members of a technological society that can keep Nature at arm's length most of the time, we have an almost autistic relationship with the Earth. We are rigid in our ways, savants when it comes to certain narrow obsessions, but dysfunctional in other regards, because we wrongly view ourselves as separate from the rest of the natural world. Convinced that Nature is something outside us, a mute and immutable thing external to us, we are unable to empathize or communicate with it.

But the Earth is speaking to us all the time. In every stone, it offers an eternal truth or good rule of thumb; in every leaf, a prototype power station; in every ecosystem, an exemplar of a healthy economy. In Aldo Leopold's words, we need to start "thinking like a mountain," awake to all the habits and inhabitants of this ancient, complicated, endlessly evolving planet.

# EPILOGUE

In 1905, John Munro Longyear, a Michigan timber and mining magnate who had made a fortune from the Proterozoic banded iron formations of the state's Upper Peninsula, was prospecting in a remote part of northern Norway with an eye toward opening a new iron range. But he needed coal for smelting, and as it happened, the nearest coalfields were on Svalbard—vestiges of an ancient tropical forest on those polar islands. He bought the mining claim from a small company based in Trondheim, set up the Arctic Coal Company, and established the town of Longyearbyen, a bit of the Wild West in the Far North. (Those unfamiliar with the origin of the name, joke that it refers to how time seems to pass in that remote place.) When Longyear found that the iron ore on the mainland was not worth extracting, the coal mines in Svalbard returned to Norwegian ownership and would remain open for more than a century. Today, some of the long adits and tunnels deep in the mountains above Longyearbyen have been repurposed as one of the world's largest seed banks (see figure 12).

The Svalbard Global Seed Vault is a library for genetic diversity, preserving the germ lines of old varieties of staple crops that may be needed as new diseases evolve or environmental changes necessitate rapid adaptation. In the event of catastrophic agricultural failures, this snow-covered mountain in the Arctic may be the bread basket of the world. Seeds are self-contained suitcases, packed and ready to travel across time even after decades of dormancy. An abandoned mine in Svalbard, the place with no official time, has become a portal into the future.

FIGURE 12. The Svalbard seed vault

Our Holocene snow day is ending now, and tomorrow's the Anthropocene. We've all enjoyed the fantasy that we can keep playing our self-absorbed and careless games—that when we choose to come inside, our supper will be waiting for us, and nothing will have changed. But no one is home to take care of us. Now we need to grow up and navigate on our own, doing our best with the Atlas of the Past to make up for so much lost time.

# APPENDIXES

## APPENDIX 1. Simplified Geologic Timescale

| | EON | ERA | PERIOD | Beginning (*millions of years ago*) | Geologic Highlights |
|---|---|---|---|---|---|
| | PHANEROZOIC | CENOZOIC | Quaternary | 3 | Human history (Holocene—10,000 years)<br>Ice Age (Pleistocene) |
| | | | Neogene | 23 | |
| | | | Paleogene | 65 | PETM (55 million years ago)<br>Mammals diversify<br>Giant birds |
| | | MESOZOIC | Cretaceous | 140 | *Dinosaur extinction*<br>Atlantic Ocean opens<br>First flowering plants |
| | | | Jurassic | 200 | *Mass extinction* |
| | | | Triassic | 250 | Age of the reptiles begins |
| | | PALEOZOIC | Permian | 290 | *Greatest mass extinction in Earth history*<br>Pangaea formed |
| | | | Carboniferous | 355 | Widespread coal swamps<br>*Mass extinction* |
| | | | Devonian | 420 | First amphibians |
| | | | Silurian | 440 | Widespread coral reefs<br>*Mass extinction* |
| | | | Ordovician | 508 | First land plants<br>First fish |
| | | | Cambrian | 541 | Modern animal phyla appear |

## APPENDIX I. *(Continued)*

| | EON | ERA | PERIOD | Beginning (*millions of years ago*) | Geologic Highlights |
|---|---|---|---|---|---|
| PRECAMBRIAN | PROTEROZOIC | NEO-PROTEROZOIC | | 565 | Ediacaran organisms |
| | | | | 800 | Snowball Earth |
| | | MESO-PROTEROZOIC | | | "Boring Billion": time of unusual climatic and geochemical stability |
| | | | | 1600 | Baraboo mountains form (Wisconsin) |
| | | PALEO-PROTEROZOIC | | 2100 | Banded iron formations are precipitated as $O_2$ accumulates in atmosphere |
| | | | | 2500 | |
| | ARCHEAN | NEOARCHEAN | | 2800 | Modern-style plate tectonics (subduction) |
| | | | | 3200 | Oldest rocks in Wisconsin |
| | | MESOARCHEAN | | | |
| | | PALEO-ARCHEAN | | 3800 | Oldest rocks in U.S. (Minnesota)<br>Earliest evidence of life (Greenland) |
| | | EOARCHEAN | | 4000 | Oldest rocks on Earth |
| | HADEAN | | | 4500 | No rocks from this period on Earth; known from meteorites, Moon rocks, and a few Australian zircon crystals |

*Note:* Intervals are not shown in proportion to duration.

## APPENDIX II. Duration and Rate of Earth Phenomena

### A. LIFESPANS

| Entity | Life Expectancy (*years*) | Limiting Processes | Chapter(s) |
|---|---|---|---|
| Our solar system | 10 billion | Sun enters red giant phase, engulfs planets | 6 |
| Total habitable time for Earth | ca. 5.5 billion (about 1.7 billion left) | Began at close of heavy meteorite bombardment period 3.8 billion years ago; will end when Sun becomes so hot that water is boiled off planet's surface | 4, 6 |
| Shield areas of continents | Up to 4 billion | Erosion | 4 |
| Ocean basin | 170 million | Ocean crust is subducted when cold and dense enough to sink into mantle | 3 |
| Mountain belt (topographic)[1] | 50–100 million | Relative rates of tectonics and erosion | 3 |
| Typical marine invertebrate species | In fossil record:[2] 10 million<br>Current species:[3] 100,000 | Sea level variation; climate change<br>Climate change; ocean acidification and anoxia | 5 |
| Typical land vertebrate species | In fossil record: 1 million<br>Current species: 10,000 | Climate change<br>Climate change; overhunting; habitat destruction | 5 |

1. Deeply eroded roots of a mountain belt with little topographic relief can survive for more billions of years.
2. May, R., Lawton, J. and Stork, N., 1995. Assessing extinction rates. In Lawton, J., and May, R. (eds.), *Extinction Rates*. Oxford: Oxford University Press, Oxford, pp. 1–24.
3. Pimm, S., et al., 1995. The future of biodiversity. *Science*, 269, 347–350.

## B. RESIDENCE AND MIXING TIMES

In geochemistry, *residence time* is the length of time a particular substance typically remains in a given site or *reservoir*. *Mixing time* is the length of time it takes such a reservoir to attain a uniform concentration of a particular substance. If the residence time is greater than the mixing time, the reservoir is well mixed with respect to that substance, and its concentration will be uniform (e.g., salt in oceans, carbon in atmosphere). If the residence time less than the mixing time, the reservoir is not well mixed with respect to that substance, and its concentration will be nonuniform (e.g., carbon in oceans).

| | Typical Value | Chapter(s) |
|---|---|---|
| **Residence Time** | | |
| Water[4] in: | | 2, 3, 6 |
| Atmosphere | 9 days | |
| Soils | 1–2 months | |
| Rivers | 2–6 months | |
| Lakes | 1–200 years | |
| Groundwater | | |
| Shallow | 10–100 years | |
| Deep | 100–10,000 years | |
| Oceans | 1000 years | |
| Glaciers | 100–800,000 years | |
| Mantle | Millions of years | |
| Carbon[5] in: | | 5 |
| Atmosphere-ocean system | 100–1000 years | |
| Soils | 25 years | |
| Land plants | 5–10 years | |
| Limestone | 10 million years | |
| Sea salt (sodium ions) | 70 million years | 3 |
| **Mixing Time** | | |
| Global ocean | ca. 1500 years | 2 |
| Troposphere (lower part of atmosphere) | 1 year | 5 |

4. University Corporation for Atmospheric Research, Center for Science Education, 2011. *The Water Cycle.* https://scied.ucar.edu/longcontent/water-cycle.
5. Kump, L., Kasting, J., and Crane, R., 1999. *The Earth System.* Englewood Cliffs, NJ: Prentice-Hall. pp. 134, 146.

## C. VELOCITIES AND RATES OF CHANGE

| | Geologic Average | Anthropocene Rate | Chapter |
|---|---|---|---|
| Plate motions | | | |
| Background rate | 1–10 cm/yr (0.4–4 in./yr) | Same | 3 |
| In earthquakes | 1 m/s (3ft/sec) | | 3 |
| Rock uplift in mountain belts | 0.1–0.5 cm/yr (0.04–2 in./yr) | Same | 3 |
| Isostatic rebound due to erosion or deglaciation | Up to 1 cm/yr (0.4 in./yr) | Same | 3 |
| Land subsidence from withdrawal of oil, gas, or groundwater | — | Up to 2 cm/yr (0.8 in./yr) | 3 |
| Erosion | 0.1 mm/yr (0.004 in./yr) (but varies with relief and climate) | ca. 1 mm/yr (0.04 in./yr)[6] | 3, 5 |
| Sea level rise | Holocene average (last 10,000 years): 0.1 mm/yr (0.004 in./yr) | Since 1900: 1.7 mm/yr (0.067 in./yr)[7] Since 1990: ca. 3.0 mm/yr (0.1 in./yr) Projected for 2100: 14 mm/yr (0.5 in./yr)[8] | 5, 6 |
| $CO_2$ emissions (as billions of tons, Gt, of carbon, C)[9] | Volcanoes: 0.2 Gt/yr | Human emissions: 10 Gt/yr | 5 |
| Increase in atmospheric $CO_2$ | Since last glacial maximum (18,000 years ago): 0.006 ppm/yr | Since 1800: 0.5 ppm/yr Since 1960: 1.5 ppm/yr Since 2000: 2.0 ppm/yr | 5 |

6. Wilkinson, B., 2005. Humans as geologic agents. *Geology*, 33, 161–164. doi:10.1130/G21108.1.
7. Church, J., and White, N., 2011. Sea level rise from the late 19th to early 21st century. *Surveys in Geophysics*, 32, 585–602. doi:10.1007/s10712-011-9119-1.
8. US Global Change Research Program, 2014. *Third National Climate Assessment*. http://www.globalchange.gov/nca3-downloads-materials.
9. Gerlach, T., 2011. Volcanic vs. anthropogenic carbon dioxide. *EOS*, 92, 201–208. doi:10.1029/2011EO240001.

## D. CYCLES AND RECURRENCE INTERVALS

| | Cycle Length | Chapter |
|---|---|---|
| Supercontinent cycle (Wilson cycle); time between assembly and breakup | ca. 500 million years | 3 |
| Milankovitch orbital cycle | | |
| Eccentricity | 96,000 and 413,000 years | 5 |
| Obliquity | 41,000 years | |
| Precession | 23,000 years | |
| Dansgaard-Oeschger cycle: (Pleistocene cooling/warming related to ocean circulation) | 1500 years | 5 |
| El Niño-Southern Oscillation (ENSO): semiperiodic alternation in location of warm water masses in Pacific Ocean; affects global weather | 3–5 years | 5 |
| Madden-Julian oscillation: repeating eastward migration of air masses over the Indian and Pacific Oceans; controls precipitation on land adjacent to both oceans | 1–3 months | 5 |
| Earth's rotation | | 4 |
| Modern | 24 hours | |
| Devonian | 22 hours | |
| Archean | 18 hours | |
| Recurrence time of supereruption at Yellowstone (last one 640,000 years ago) | ca. 700,000 years | 2, 3 |
| Recurrence time of M9 earthquakes on Cascadia subduction zone (last in 1700) | 200–800 years | 3 |
| Global earthquake recurrence time (long-term averages) | | 3 |
| Magnitude 9 | 10 years | |
| Magnitude 8 | 1 year | |
| Magnitude 7 | 1 month | |
| Magnitude 6 | 1 week | |

# APPENDIX III: Environmental Crises in Earth's History: Causes and Consequences

| EVENT[1] | Extinction Severity[2] | C Cycle Perturbation: Volcanic/ Tectonic | C Cycle Perturbation: Biogenic $(\Delta\delta^{13}C)$[3] | Climate Change | Sea Level | Ocean Acidity | Ocean Anoxia | Ozone Destruction | Aftermath/ Legacy |
|---|---|---|---|---|---|---|---|---|---|
| Snowball Earth 750–570 Ma | Unknown—likely severe | Initial cooling: C sequestration > volcanic emissions | Possibly ended by methane hydrate release $(\Delta\delta^{13}C = -10)$[4] | Extreme cold, then extreme warmth | Very low to high | | | | Ediacaran fauna, then Cambrian explosion |
| End-Ordovician extinction (#2) 440 Ma | 57% of genera 86% of species | Probably some type of C cycle disturbances, but not well constrained | | Abrupt ice age followed by rapid warming | High to low to high | | Yes | | Cambrian organisms (e.g., trilobites) decimated |
| Late Devonian extinction (#4) 365 Ma | 35% of genera 75% of species | | Biogenic C burial > decomposition $(\Delta\delta^{13}C = \text{ca. } +4)$[5] | Abrupt cooling | High to low | | Yes | | Marine filter-feeders diversify |
| End-Permian extinction (#1) 250 Ma | 56% of genera 95% of species | Siberian Traps (flood basalts) | Methane hydrates and/ or burning coal seams $(\Delta\delta^{13}C = -8)$[6] | Cold to extremely warm | Low to high | Yes | Yes | Yes—by volcanic gases | Permanent ecosystem reorganization; low $O_2$ for >1 million years |
| End-Triassic (#3) 200 Ma | 47% of genera 80% of species | Central Atlantic flood basalts | $(\Delta\delta^{13}C = -3)$[7] | Hot and dry | | Yes | | | Dinosaurs diversify |

| | | | | | | | | | |
|---|---|---|---|---|---|---|---|---|---|
| End-Cretaceous (#4) 65 Ma | 40% of genera 76% of species | Meteorite impact releases $CO_2$ from carbonates Deccan Traps | ($\Delta\delta^{13}C = -1$) | Short cold spell (ash, $SO_2$), then long warm period ($CO_2$) | | Yes | | Maybe—chlorine from seawater vaporized in impact? | Dinosaurs vanish (except birds); mammals diversify |
| Paleocene-Eocene Thermal Max. 55 Ma | Deep-ocean foraminifera hit hard | North Atlantic flood basalts | Methane hydrates and/or burning coal seams ($\Delta\delta^{13}C = -3$)[8] | Warming spike | Rapid rise | Yes | | | No ice; major land and deep-sea ecosystem changes |
| Anthropocene | Extinction rates 100–100X background | | Fossil fuel combustion ($\Delta\delta^{13}C = -2$)[9] | Rapid warming | Rapid rise | Yes | Yes | Yes | ? ? |

1. Time of events is given in millions of years before the present (Ma). For the five great mass extinctions, the rank in severity is indicated in parentheses.
2. Values from Barnosky, A., et al., 2011. Has the sixth mass extinction already arrived? *Nature*, 471, 51–57. doi: 10.1038/nature09678.
3. The value $\Delta\delta^{13}C$ is a measure of the change in the ratio of stable carbon isotopes ($^{13}C$ and $^{12}C$) in seawater from a background value—and thus a measure of the severity of disturbance to the carbon cycle. $\delta^{13}C$ ("delta C-13") is defined as $[(^{13}C/^{12}C$ calcite sample $-$ $^{13}C/^{12}C$ calcite standard$)/$ $^{13}C/^{12}C$ calcite standard$] \times 1000$. (The factor of 1000 is used so that the differences have integer values; variations in the $^{13}C/^{12}C$ ratio are measured in parts per thousand). The value $\Delta\delta^{13}C$ ("delta delta C-13") means a change in the $\delta^{13}C$ value over a particular interval of time. A shift to more negative values indicates release of biogenic (photosynthetically fixed) carbon. A shift to more positive values indicates organic carbon burial and/or the dominance of volcanic $CO_2$ over biogenic $CO_2$ emissions.
4. Snowballearth.org.
5. Buggish, W., and Joachimski, M., 2006. Carbon isotope stratigraphy of the Devonian of Central and Southern Europe, *Palaeogeography, Palaeoclimatology, Palaeoecology*, 240, 68–88.
6. Erwin, D. H., 1994. The Permo-Triassic extinction. *Nature*, 367, 231–236, doi:10.1038/367231a0.
7. Schoene, B., et al., 2010. Correlating the end-Triassic mass extinction with flood basalt volcanism at the 100,000 year level. *Geology*, 38, 387–390. doi:10.1130/G30683.1.
8. Tipple, B., et al., 2011. Coupled high-resolution marine and terrestrial records of carbon and hydrologic cycles variations during the Paleocene-Eocene Thermal Maximum (PETM). *Earth and Planetary Science Letters*, 311, 82–92. doi:10.1016/j.epsl.2011.08.045.
9. Friedli, D., et al., 1986. Ice core record of the $^{13}C/^{12}C$ ratio of atmospheric $CO_2$ in the last two centuries. *Nature*, 324, 237–238.

# NOTES

## 1. A CALL FOR TIMEFULNESS

1. Descartes, R., 1641, translated by Michael Moriarty, 2008. *Meditations on First Philosophy, with Selections from the Objections and Replies*. Oxford: Oxford World's Classics, p. 16.
2. Haldane is supposed to have said this when asked what could cause him to abandon his certitude about evolution. The memorable quote has been cited many times, but its origin is not clear.
3. Barker, D., and Bearce, D., 2012. End-times theology, the shadow of the future, and public resistance to addressing climate change. *Political Research Quarterly*, 66, 267–279. doi:0.1177/1065912912442243.
4. Baumol, W., and Bowen, W., 1966. *Performing Arts—The Economic Dilemma: A Study of Problems Common to Theater, Opera, Music, and Dance*. New York: Twentieth Century Fund, 582 pp.
5. Theoretical physicist Lee Smolin is a minority voice chiding his discipline for what he calls the systematic "expulsion of time." Smolin, L., 2013, *Time Reborn*, Boston: Houghton Mifflin Harcourt, 352 pp.
6. Including Steven Levitt and Stephen Dubner in Chapter 5 of *Superfreakonomics: Global Cooling, Patriotic Prostitutes, and Why Suicide Bombers Should Buy Life Insurance*. 2010. New York: William Morrow, 320 pp.

## 2. AN ATLAS OF TIME

1. McPhee, J., 1981. *Basin and Range*. New York: Farrar, Strauss and Giroux, p. 20.
2. It should be noted that some non-Western cultures had prescientific concepts of "Deep Time." For example, Hinduism and Buddhism share the concept of *kalpa*, a Sanskrit word for a cosmological eon—an interval of time far longer than human experience and memory. Other cultures outside the Abrahamic traditions likely had similar ideas about the antiquity of the Universe. But in Europe, where modern geologic thinking began, biblical doctrine was long a barrier to scientific understanding.
3. While this number was not an accurate determination of the age of the Earth, it is not without meaning; it is close to the modern estimate of the average time an atom of sodium remains in the sea (its *residence time*) before being removed via sea spray or precipitation of rock salt. See Appendix II for characteristic residence times of other geologic "commodities."
4. Thomson, W., (Lord Kelvin) 1872. President's Address. *Report of the Forty-First Meeting of the British Association for the Advancement of Science*, Edinburgh,

pp. lxxiv-cv. Reprinted in Kelvin, 1894, *Popular Lectures and Addresses*, vol. 2. London: Macmillan, pp. 132–205.

5. For a lively, readable biography of Arthur Holmes, see Cherry Lewis, 2000, *The Dating Game: One Man's Search for the Age of the Earth*. Cambridge: Cambridge University Press.
6. The Rutherford-Soddy law, the mathematical description of radioactive decay, is $dP/dt = -\lambda P$, where $P$ is the number of atoms of the parent isotope at any given time, $dP/dt$ is the rate of decay, and $\lambda$ is the decay constant for that isotope. The relationship between the half-life $t_1/_2$ and the decay constant is $t_1/_2 = \ln 2/\lambda$, or $0.693/\lambda$. In about 10 mathematical steps, it is possible to derive from Rutherford's law an equation—the Age Equation—that expresses the age of a mineral (time elapsed since crystallization, $t$) as a function of the daughter/parent ratio $D/P$ and the decay constant $\lambda$. It is simply: $t = 1/\lambda\ [\ln (D/P + 1)]$.
7. International Commission on Stratigraphy: http://www.stratigraphy.org/index.php/ics-gssps.
8. An audio interview with Nier about his work before and during the Manhattan Project can be heard at http://manhattanprojectvoices.org/oral-histories/alfred-niers-interview-part-1.
9. It should be acknowledged that a Russian geochemist, E. K. Gerling, carried out a very similar calculation at virtually the same time as Holmes and obtained an age of 3.1 billion years. But his work was not known in the West until much later. See Dalrymple, G. B., 2001. The age of the Earth in the twentieth century: A problem (mostly) solved. In Lewis, C. and Knell, S., *The Age of the Earth from 4004 BC to AD 2002*. Geological Society of London Special Publication 190, 205–221.
10. Brush, S., 2001. Is the Earth too old? The impact of geochronology on cosmology, 1929–1952. In Lewis, C., and Knell, S., *The Age of the Earth from 4004 BC to AD 2002*. Geological Society of London Special Publication 190, 157–175.
11. Patterson, C., 1956. Age of meteorites and the Earth. *Geochimica et Cosmochimica Acta*, 10, 230–277. doi:10.1016/0016-7037(56)90036-9.
12. Coleman, D., Mills, R., and Zimmerer, M., 2016. The pace of plutonism. *Elements*, 12, 97–102. doi:10.2113/gselements.12.2.97.
13. Gebbie, G., and Huybers, P., 2012. The mean age of ocean waters inferred from radiocarbon observations: Sensitivity to surface sources and accounting for mixing histories. *Journal of Physical Oceanography*, 42, 291- 305. doi:10.1175/JPO-D-11-043.1.
14. Suess H., 1955. Radiocarbon concentration in modern wood. *Science*, 122, 415–417.
15. For a lyrical account of the geology of the Apennines, see Walter Alvarez, 2008. *In the Mountains of St Francis*. New York: WW Norton.
16. Genge, M., et al., 2016. An urban collection of modern-day large micrometeorites: Evidence for variations in the extraterrestrial dust flux through the Quaternary. *Geology*, 45, 119–121. doi:10.1130/G38352.1.
17. Swisher et al., 1992. Coeval $^{40}Ar/^{39}Ar$ ages of 65.0 million years ago from Chicxulub Crater melt rock and Cretaceous-Tertiary boundary tektites, *Science*, 257, 954–958.

18. Wilde, S., Valley, J., Peck, W., and Graham, C., 2001. Evidence from detrital zircons for the existence of continental crust and oceans on the Earth 4.4 Gyr ago. *Nature,* 409, 175–178. doi:10.1038/35051550.

## 3. THE PACE OF THE EARTH

1. The lack of detailed information about seafloor topography was underscored by the search for the wreckage of Malaysian Airlines Flight 370, which disappeared somewhere in the Indian Ocean in March 2014. In 2016, an international geophysical team conducted echo soundings along a swath 100 miles wide and 1500 miles long in an area about 1000 miles west of Australia, revealing many previously unknown fracture zones, escarpments, landslides, and volcanic centers, but no trace of the lost aircraft. See Picard, K., Brooke, B., and Coffin, M., 2017. Geological insights from Malaysia Airlines Flight MH370 search. *EOS, Transactions of the American Geophysical Union*, 98. https://doi.org/10.1029/2017EO069015.
2. A marvelous biography of Marie Tharp is *Soundings: The Story of the Remarkable Woman who Mapped the Ocean Floor*, by Hali Felt (2012). New York: Henry Holt, 368 pp.
3. Vine, F., and Matthews, D., 1963. Magnetic anomalies over mid-ocean ridges. *Nature,* 199, 947–950.
4. East Pacific Rise Study Group, 1981. Crustal processes of the mid-ocean ridge, *Science*, 213, 31–40.
5. Gondwanaland, which included India, Africa, South America, Australia, and Antarctica, was first hypothesized and named in the 1880s by the Austrian geologist Edward Suess based on similarities in the fossils, rock strata, and ancient mountain ranges of the southern landmasses. The name was later used by the German meteorologist Alfred Wegener in his 1915 treatise *Origin of Continents and Oceans*, which made a strong case for continental drift a half-century before the discovery of seafloor spreading and development of plate tectonic theory.
6. Ruskin, J., 1860. *Modern Painters*, vol. 4: *Of Mountain Beauty*, p. 196–197. Available through Project Gutenberg: http://www.gutenberg.org/files/31623/31623-h/31623-h.htm.
7. Liang, S., et al., 2013. Three-dimensional velocity field of present-day crustal motion of the Tibetan Plateau derived from GPS measurements. *Journal of Geophysical Research: Solid Earth*, 118, 5722–5732. doi:10.1002/2013JB010503.
8. Van der Beek, P., et al., 2006. Late Miocene—Recent exhumation of the central Himalaya and recycling in the foreland basin assessed by apatite fission-track thermochronology of Siwalik sediments, Nepal. *Basin Research*, 18, 413–434.
9. Clift, P. D., et al., 2001. Development of the Indus Fan and its significance for the erosional history of the Western Himalaya and Karakoram. *Geological Society of America Bulletin*, 113, 1039–1051.
10. Einsele, G., Ratschbacher, L., and Wetzel, A., 1996. The Himalaya-Bengal fan denudation-accumulation system during the past 20 Ma. *Journal of Geology*, 104, 163–184. doi:10.1086/629812.

11. Curray, J., 1994. Sediment volume and mass beneath the Bay of Bengal. *Earth and Planetary Science Letters*, 125, 371–383.
12. Based on the Plateau area of 2.6 million $km^2$ and average elevation of 4.5 km.
13. Seong, Y., et al., 2008. Rates of fluvial bedrock incision within an actively uplifting orogen: Central Karakoram Mountains, northern Pakistan, *Geomorphology*, 97, 274–286. doi:10.1016/j.geomorph.2007.08.011.
14. Davies, N., and Gibling M., 2010. Cambrian to Devonian evolution of alluvial systems: The sedimentological impact of the earliest land plants. *Earth Science Reviews*, 98, 171–200. doi:10.1016/j.earscirev.2009.11.002.
15. Brown, A. G., et al., 2013. The Anthropocene: Is there a geomorphological case? *Earth Surface Processes and Landforms*, 38, 431–434. doi:10.1002/esp.3368.
16. Lim, J., and Marshall, C., 2017. The true tempo of evolutionary radiation and decline revealed on the Hawaiian archipelago. *Nature*, 543, 710–713. doi:10.1038/nature21675.
17. For a survey of the many feedbacks between topography, climate, and erosion see Brandon, M., and Pinter, N., How erosion builds mountains, *Scientific American*, July 2005.
18. In central Sweden, postglacial rebound rates are on the order of 0.6 cm (0.25 in.) per year—fast enough that settlements that were seaports in Viking times now lie on inland lakes. Neighboring Finland has laws governing who owns new coastal land that emerges from the sea; these may become moot, however, if sea level rise outpaces isostatic uplift.
19. Champagnac, J., et al., 2009. Erosion-driven uplift of the modern Central Alps. *Tectonophysics*, 474, 236–249. doi:10.1016/j.tecto.2009.02.024.
20. Darwin, C., 1839. *Voyage of the Beagle*, chap. 14.
21. Stein, S., and Okal, E., 2005. Speed and size of the Sumatra earthquake. *Nature*, 434, 581–582. doi:10.1038/434581a.
22. Ben-Naim, E., Daub, E., and Johnson, P., 2013. Recurrence statistics of great earthquakes. *Geophysical Research Letters*, 40, 3021–3025, doi:10.1002/grl.50605.
23. Houston, H., et al., 2011. Rapid tremor reversals in Cascadia generated by a weakened plate interface. *Nature Geoscience*, 4, 404–408. doi:10.1038/NGEO1157.
24. Brudzinksi, M., and Allen, R., 2007. Segmentation in episodic tremor and slip all along Cascadia. *Geology*, 35, 907–910. doi:10.1130/G23740A.1.
25. Yamashita, Y., et al., 2015. Migrating tremor off southern Kyushu as evidence for slow slip of a shallow subduction interface. *Science*, 348, 676–679. doi:10.1126/science.aaa4242.
26. Booth, A., Roering, J., and Rempel, A., 2013. Topographic signatures and a general transport law for deep-seated landslides in a landscape evolution model. *Journal of Geophysical Research: Earth Surface*, 118, 603–624. doi:10.1002/jgrf.20051.
27. Parker, R., et al., 2011. Mass wasting triggered by the 2008 Wenchuan earthquake is greater than orogenic growth. *Nature Geoscience*, 4, 449–452.

28. Ramalho, R., et al., 2015. Hazard potential of volcanic flank collapses raised by new megatsunami evidence. *Science Advances*, 1, e1500456. doi:10.1126/sciadv.1500456.
29. Aranov, E., and Anders, M., 2005. Hot water: A solution to the Heart Mountain detachment problem? *Geology*, 34, 165–168. doi:10.1130/G22027.1; Craddock, J., Geary, J. and Malone, D., 2012. Vertical injectites of detachment carbonate ultracataclasite at White Mountain, Heart Mountain detachment, Wyoming. *Geology*, 41, 463–466. doi:10.1130/G32734.1.
30. Ross, M., McGlynn, B., and Bernhardt, E., 2016. Deep impact: Effects of mountaintop mining on surface topography, bedrock structure and downstream waters. *Environmental Science and Technology*, 50, 2064–2074. doi:10.1021/acs.est.5b04532.
31. Wilkinson, B., 2005. Humans as geologic agents: A deep-time perspective. *Geology*, 33, 161–164. doi:10.1130/G21108.1.
32. Hurst, M., et al., 2016. Recent acceleration in coastal cliff retreat rates on the south coast of Great Britain. *Proceedings of the National Academy of Sciences*, 113, 13336–13341, doi:10.1073/pnas.1613044113.
33. Stanley, J.-D., and Clemente, P., 2017. Increased land subsidence and sea-level rise are submerging Egypt's Nile Delta coastal margin. *GSA Today*, 27, 4–11. doi:10.1130/GSATG312A.1.
34. Morton, R., Bernier, J., and Barras, J., 2006. Evidence of regional subsidence and associated interior wetland loss induced by hydrocarbon production, Gulf Coast region, USA. *Environmental Geology*, 50, 261–274.
35. According to a U.S. Geological Survey report, seismic risk from human-induced earthquakes in Oklahoma in 2017 equaled that of natural earthquakes in California: Peterson, M., et al., 2017. One-year seismic-hazard risk forecast for the central and eastern Unites States from induced and natural earthquakes. *Seismological Research Letters*, 88, 772–783. doi:10.1785/0220170005.

## 4. CHANGES IN THE AIR

1. Marchis, S., et al., 2016. Widespread mixing and burial of Earth's Hadean crust by asteroid impacts. *Nature*, 511, 578–582. doi:10.1038/nature13539.
2. Williams, G., 2000. Geological constraints on the Precambrian history of Earth's rotation and the Moon's orbit. *Reviews of Geophysics*, 38, 37–59. doi:10.1029/1999RG900016.
3. Sagan, C., and Mullen, G., 1972. Earth and Mars: Evolution of atmospheres and surface temperatures. *Science*, 177, 52–56.
4. Mojzsis, S. J., et al., 1996. Evidence for life on Earth before 3800 million years ago. *Nature*, 384, 55–59. doi:10.1038/384055a0.
5. van Zuilen, M., Lepland, A., and Arrhenius, G, 2002. Reassessing the evidence for the earliest traces of life. *Nature*, 418, 627–630. doi:10.1038/nature00934.
6. Whitehouse, M., Myers, J., and Fedo, C., 2009. The Akilia Controversy: Field, structural and geochronological evidence questions interpretations of

>3.8 Ga life in SW Greenland. *Journal of the Geological Society*, 166, 335–348. doi:10.1144/0016-76492008-070.

7. Westall, F., and Folk, R., 2003. Exogenous carbonaceous microstructures in Early Archean cherts and BIFs from the Isua Greenstone Belt: Implications for the search for life in ancient rocks. *Precambrian Research* 126, 313–330.
8. Van Kranendonk, M., Philippot, P., Lepot, K., Bodorkos, S. & Pirajno, F., 2008. Geological setting of Earth's oldest fossils in the c. 3.5 Ga Dresser Formation, Pilbara craton, Western Australia. *Precambrian Research* 167, 93–124.
9. Nutman, A., Bennett, V., Friend, C., Van Kranendonk, M., and Chivas, A., 2016. *Nature*, 537 http://dx.doi.org/10.1038/nature19355.
10. Watson, Traci. 3.7 billion year old fossil makes life on Mars less of a long shot, USA Today, 31 August 2016. http://www.usatoday.com/story/news/2016/08/31/37-billion-year-old-fossil-makes-life-mars-less-long-shot/89647646/.
11. Zerkle, A., et al., 2017. Onset of the aerobic nitrogen cycle during the Great Oxidation Event. *Nature*, doi:10.1038/nature20826.
12. Kump, L. and Barley, M., 2007. Increased subaerial volcanism and the rise of oxygen 2.5 billion years ago. *Nature*, 448, 1033–1036. doi:10.1038/nature06058.
13. Johnson, T., et al., 2014. Delamination and recycling of Archean crust caused by gravity instabilities. *Nature Geoscience*, 7, 47–52. doi:10.1038/ngeo2019.
14. Lyons, T., Reinhard, C., and Planavsky, N., 2014. The rise of oxygen in Earth's early ocean and atmosphere. *Nature*, 307, 506–511. doi:10.1038/nature13068.
15. Planavsky, N., et al., 2014. Low mid-Proterozoic atmospheric oxygen levels and the delayed rise of animals. *Science*, 346, 635–638. doi:10.1126/science.1258410.
16. Reinhard, C., et al., 2016. Evolution of the global phosphorus cycle, *Nature*, doi:10.1038/nature20772.
17. Wolf, E., and Toon, O., 2015. Delayed onset of runaway and moist greenhouse climates for Earth. *Geophysical Research Letters*, 41, 167–172. doi:10.1002/2013GL058376. The good news is that this study extended the habitable period from the truly depressing estimates of 170–650 million years!
18. Planavsky, N., et al., 2010. The evolution of the marine phosphate reservoir. *Nature*, 467, 1088–1090.
19. Erwin, D., et al., 2011. The Cambrian conundrum: Early divergence and later ecological success in the early history of animals. *Science*, 334, 1091–1097. doi:10.1126/science.1206375.
20. Kelvin's phrase, in a letter to John Phillips. Quoted in Morrell, J., 2001. The age of the Earth in the twentieth century: A problem (mostly) solved. In Lewis, C., and Knell, S., *The Age of the Earth from 4004 BC to AD 2002*. Geological Society of London Special Publication 190, 85–90.
21. McCallum, M., 2007. Amphibian decline or extinction? Current declines dwarf background extinction rate. *Journal of Herpetology*, 41, 483–491. doi:10.1670/0022–1511.
22. Raup, D., and Sepkoski, J., 1984. Periodicity of extinctions in the geologic past. *Proceedings of the National Academy of Sciences*, 81, 801–805.

23. Whitman, W., Coleman, D., and Wiebe, W., 1998. Prokaryotes: The unseen majority. *Proceedings of the National Academy of Sciences*, 95, 6578–6583.

## 5. GREAT ACCELERATIONS

1. Cooper, K., and Kent, A., 2014. Rapid remobilization of magmatic crystals kept in cold storage. *Nature*, 506, 480–483. doi:10.1038/nature12991.
2. Webber, K., et al., 1999. Cooling rates and crystallization dynamics of shallow level pegmatite-aplite dikes, San Diego County, California. *American Mineralogist*, 84, 718–717.
3. Zalasiewicz, J., et al., 2008. Are we now living in the Anthropocene? *GSA Today*, 18(2), 4–8. doi:10.1130/GSAT01802A.1.
4. Lambeck, K., et al., 2014. Sea level and global ice volumes from the Last Glacial Maximum to the Holocene. *Proceedings of the National Academy of Sciences*, 111, 15296–15303. doi:10.1073/pnas.1411762111.
5. Center for Biological Diversity, http://www.biologicaldiversity.org/programs/biodiversity/elements_of_biodiversity/extinction_crisis/.
6. Gerlach, T., 2011. Volcanic vs. anthropogenic carbon dioxide. *Eos, Transactions, American Geophysical Union*, 92, 201–203.
7. Rockström, J., et al., 2009. A safe operating space for humanity. *Nature*, 461, 472–475. doi:10.1038/461472a.
8. Haberl, H., et al., 2007. Quantifying and mapping the human appropriation of net primary production in Earth's terrestrial ecosystem. *Proceedings of the National Academy of Sciences*, 104, 12942–12947. doi:10.1073/pnas0704243104.
9. Walker, M., et al., 2009. Formal definition and dating of the GSSP (Global Stratotype Section and Point) for the base of the Holocene using the Greenland NGRIP ice core, and selected auxiliary records. *Journal of Quaternary Science*, 24, 3–17. doi:10.1002/jqs.1227.
10. Thompson, L., et al., 2013. Annually resolved ice core records of tropical climate variability over the past 1800 Years. *Science*, 340, 945–950. doi:10.1126/science.123421.
11. Zhang, D., et al., 2011. The causality analysis of climate change and large-scale human crisis. *Proceedings of the National Academy of Sciences*, 108, 17296–17301. doi:10.1073/pnas.1104268108.
12. Hsiang, S., Burke, M., and Michel, E., 2013. Quantifying the influence of climate on human conflict, *Science*, 341, 1212–1228. doi:10.1126/science.1235367.
13. Milly, P., et al., 2008. Stationarity is dead: Whither water management? *Science*, 319, 573–574. doi:10.1126/science.1151915.
14. Alley, R., 2000. *The Two-Mile Time Machine: Ice Cores, Abrupt Climate Change, and our Future*. Princeton, NJ: Princeton University Press, p. 126.
15. Berger, A., 2012. A brief history of the astronomical theories of paleoclimate. *In* Berger A., Mesinger F., and Sijacki, D. (eds.), *Climate Change*. New York: Springer, p. 107–128. doi:10.1007/978-3-7091-0973-1_8.

16. Arrhenius, S., 1896. On the influence of carbonic acid in the air upon the temperature of the ground. *Philosophical Magazine and Journal of Science*, ser. 5, vol. 41, 237–276.
17. Hays, J., Imbrie, J., and Shackleton, N., 1976. Variations in the Earth's orbit: Pacemaker of the ice ages. *Science*, 194, 1121–1132.
18. A provocative segment in Neil deGrasse Tyson's 2014 excellent TV series *Cosmos* depicts a city as it would look if $CO_2$ were a purple gas. Carbon emissions would then be considered a public menace.
19. The relative amount of $^{13}C$ and $^{12}C$ in a geologic sample is typically given in terms of the deviation of the $^{13}C/^{12}C$ ratio in a given rock (usually limestone) from an international standard (a "reference" piece of calcite). This deviation is called $\delta^{13}C$ (delta C-13) and defined as

$$[(^{13}C/^{12}C \text{ sample} - {}^{13}C/^{12}C \text{ standard}) / {}^{13}C/^{12}C \text{ standard}] \times 1000.$$

(The factor of 1000 is used so that the differences have integer values; variations in the $^{13}C/^{12}C$ ratio are measured in parts per thousand). The change in the $\delta^{13}C$ value in rocks over some period of time—denoted $\Delta\delta^{13}C$ (delta delta C-13)—is a measure of the severity of disturbance to the carbon cycle. A negative value of $\Delta\delta^{13}C$ indicates release of biogenic (photosynthetically fixed) carbon. A positive value indicates organic carbon burial and/or the dominance of volcanic $CO_2$ over biogenic $CO_2$ emissions. See also Appendix III.

20. McInerney, F., and Wing, S., 2011. The Paleocene-Eocene Thermal Maximum: A perturbation of carbon cycle, climate, and biosphere with implications for the future. *Annual Reviews of Earth and Planetary Sciences*, 39, 489–516.
21. Union of Concerned Scientists. Environmental impacts of natural gas, https://www.ucsusa.org/clean-energy/coal-and-other-fossil-fuels/environmental-impacts-of-natural-gas.
22. Ruben, E., Davidson, J., and Herzog, H., 2015. The cost of $CO_2$ capture and storage. *International Journal of Greenhouse Gas Control*. doi:10.1016/j.ijggc.2015.05.018.
23. American Physical Society, 2011. Direct air capture of $CO_2$ with chemicals. https://www.aps.org/policy/reports/assessments/.
24. Stephenson, N. L., et al., 2014. Rate of tree carbon accumulation increases continuously with tree size. *Nature*, 507, 90–93. doi:10.1038/nature12914.
25. Venton, D., 2016. Can bioenergy with carbon capture and storage make an impact? *Proceedings of the National Academy of Sciences*, 47, 13260–13262. doi:10.1073/pnas.1617583113.
26. American Society for Microbiology, 2017. Colloquium Report: *Microbes and Climate Change*. https://www.asm.org/index.php/colloquium-reports/item/4479-microbes-and-climate-change.
27. Keleman, P., and Metter, J., 2008. In situ carbonation of peridotite for $CO_2$ storage. *Proceedings of the National Academy of Sciences*, 105, 17295–17300. doi:101073/pnas.0805794105.

28. Hamilton, Clive, 2013. *Earthmasters: The Dawn of the Age of Climate Engineering*. New Haven, CT: Yale University Press.
29. Smith, C. J., et al., 2017. Impacts of stratospheric sulfate geoengineering on global solar photovoltaic and concentrating solar power resource. *Journal of Applied Meteorology and Climatology*, 56, 1484–1497. doi:10.1175/JAMC-D-16–0298.1.
30. Tilmes, S., et al., 2013. The hydrological impact of geoengineering in the Geoengineering Model Intercomparison Project (GeoMIP). *Journal of Geophysical Research: Atmospheres*, 118, 11036011958. doi:10.1002/jgrd.50868.
31. Keith, D., 2013. *A Case for Climate Engineering*. Cambridge, MA: MIT Press.

## 6. TIMEFULNESS, UTOPIAN AND SCIENTIFIC

1. For Packer cognoscenti: Desmond Bishop.
2. Wisconsin Department of Natural Resources, Winnebago System Sturgeon Spearing, http://dnr.wi.gov/topic/fishing/sturgeon/sturgeonlakewinnebago.html.
3. LaTour, B., 1993. *We Have Never Been Modern*. Cambridge, MA: Harvard University Press, p. 68.
4. Shulman, E., 2014. *Rethinking the Buddha: Early Buddhist Philosophy as Meditative Perception*, Cambridge: Cambridge University Press, p. 114.
5. A thousand years later, another Scandinavian, the Danish theologian and philosopher Søren Kierkegaard (who would certainly have denied any lingering Viking influences) posited the complementary idea that "the future signifies more than the present and the past; for the future is in a sense the whole of which the past is a part." [Kierkegaard, 1844, *The Concept of Dread*].
6. Bauschatz, P., 1982. *The Well and the Tree*. Amherst: University of Massachusetts Press.
7. Bergquist, L., 2016. Brad Schimel opinion narrows DNR powers on high-capacity wells. *Milwaukee Journal Sentinel*, 16 May 2016, http://archive.jsonline.com/news/statepolitics/brad-schimel-opinion-narrows-dnr-powers-on-high-capacity-wells-brad-schimel-opinion-narrows-dnr-powe-378900981.html.
8. Wieseltier, L., 2015. *Among the Disrupted*, *New York Times Book Review*, 7 Jan. 2015.
9. The full text of the Great Law can be found at http://www.indigenouspeople.net/iroqcon.htm.
10. Scheffler, S., 2016. *Death and the Afterlife*. Oxford: Oxford University Press, p. 43.
11. Hauser, O., et al., 2014. Cooperating with the future. *Nature*, 511, 220–223. doi:10.1038/nature13530.
12. Hardin, G., 1969. The tragedy of the commons. *Science*, 162, 1243–1248.
13. Sussman, R., 2014. *The Oldest Living Things in the World*. Chicago: University of Chicago Press.
14. Smith, R., 2014. On Kawara, artist who found elegance in every day dies at 81. *New York Times*, 15 July 2014. https://www.nytimes.com/2014/07/16/arts

/design/on-kawara-conceptual-artist-who-found-elegance-in-every-day-dies-at-81.html.

15. John Cage Orgelprojekt Halberstadt. http://www.aslsp.org/de/.
16. The Long Now Foundation. http://longnow.org/clock/.
17. Feder, T., 2012. Time for the future. *Physics Today*, 65(3), 28.
18. Osnos, E., 2017. Survival of the richest. *New Yorker*, 30 January 2017.
19. Vizenor, G., 2008. *Survivance: Narratives of Native Presence*. Lincoln: University of Nebraska Press.
20. Loew, P., 2014. *Seventh Generation Earth Ethics: Native Voices of Wisconsin*. Madison: University of Wisconsin Press.
21. Wolf, E., and Toon, O., 2015. The evolution of habitable climates under the brightening Sun. *Journal of Geophysical Research: Atmospheres*, 120, 5775–5794. doi:10.1002/2015JD023302.
22. http://www.scotese.com/future2.htm. See also Broad, W., 2007, Dance of the continents. *New York Times*, 9 January 2007. http://www.nytimes.com/2007/01/09/science/20070109_PALEO_GRAPHIC.html?mcubz=2.
23. Vonnegut was interviewed in 2005 by David Brancaccio on *PBS Now*. http://www.pbs.org/now/transcript/transcriptNOW140_full.html.
24. Tracy, J., Hart, H., and Martens, J., 2011. Death and science: The existential underpinnings of belief in intelligent design and discomfort with evolution. *PloSONE* 6: e17349. doi:10.1371/journal.pone.0017349. http://www.plosone.org/article/info%3Adoi%2F10.1371%2Fjournal.pone.0017349.
25. Dobzhansky, T., 1973. Nothing in biology makes sense except in the light of evolution. *American Biology Teacher*, 35(3), 125–129. It should be noted that Dobzhansky was a theist and devout member of the Eastern Orthodox Church who saw no conflict between his work in evolutionary biology and his belief in God.
26. Smolin L., 2014. Time, laws, and the future of cosmology. *Physics Today*, 67(3), 38–43.
27. Freud, S., 1929, translated by James Strachey, 1961. *Civilization and Its Discontents*. New York: W.W. Norton, p. 15–19.
28. Durkheim, É., 1912. *The Elementary Forms of the Religious Life*. Translated by K. Fields, New York: Free Press (1995), p. 228.

# INDEX


Acasta gneiss, 98, 197, 110
acid rain, 120–21, 156
Africa, 53, 55, 88
Agassiz, Louis, 135
Agassiz, Glacial Lake, 135
age of the Earth, 23, 28–30, 40–47
albedo effect, 112, 142
Alley, Richard, 134
Alps, 39, 73, 74, 83, 135, 172
Aleutian Islands, 72
Alvarez, Luis and Walter, 54–56, 119
amphibians, 118, 121
Andes, 72, 131
animals, origin of 115–6
Angkor Kingdom, 132
Anning, Mary, 27
anoxia, oceanic, 123–4, 129, 153
Antarctica, 2,55, 138, 168
Anthropocene, 91, 128–131, 133, 167, 177
Apennines, Italian, 55
apocalypticism, 11–12, 119
Appalachian Mountains, 3, 38, 89–90
Archean Eon, 98–103
archeology, 51, 53
Arctic, 2–5, 81, 93–95, 141–2, 180
Argon-argon dating, 54–57
Argonne National Laboratory, 45
Arhennius, Svante, 138
ash, volcanic, 37–38, 53
Asia, 73, 74, 132, 156
asthenosphere, 83
astrobiology, 14
Atlantic Ocean, 62, 68, 69, 143, 147
Atwood, Margaret, 168
Australia, 58–59, 101–2, 114, 167

Bangladesh, 77
baobab trees, 167
Baraboo Hills, Wisconsin, 78
basalt, 62–63, 70–73, 99, 103, 121
Baumol's "disease," 12
Beagle, Voyage of, 26, 84
Becquerel, Henri, 33
Bengal Fan (Indian Ocean), 77, 89
Big Bang theory, 43
biogeochemical cycles, 81–83, 97, 128, 148, 187; disruptions of, in mass extinctions, 123–5; human perturbations of, 129, 153, 155; in Proterozoic time, 105–9. *See also* carbon cycle.
biostratigraphy, 36
Blackmore, Susan, 175
black smokers, 71
"Boring Billion" (interval in Proterozoic time), 108–11, 128
brachiopods, 114
Brahmaputra River, 77, 89
Brazil, 69
bristlecone pine, 167
British Columbia, 87, 153
Buddhism, 162
Bunin, Rabbi Simcha, 177

$^{14}C$ dating, 50–52, 59
Cage, John, 168–9
calcite, 82–83, 114, 123, 146, 153–4
Caledonides, 3, 10, 73
California, 154
California Institute of Technology, 45
Cambrian explosion, 114–5, 123
Cambrian Period, 28, 29, 40, 58, 114–5
Cambridge University, 33
Canadian Shield, 53, 57–58, 98, 110
Cape Verde Islands, 88
carbon capture and storage, 148–153
carbon cycle, on geologic timescales, 81–83, 106, 141–2; 172; human perturbations of, 143–54; and mass extinctions, 123–4

carbon dating. *See* $^{14}$C dating
carbon dioxide in atmosphere, 81–83, 96, 121–4, 129, 138, 141–55
carbon market or tax, 149, 150
carbon, stable isotopes of, 100, 139, 146
Carboniferous Period, 40
Caribbean Sea, 56
Carnegie, Andrew, 103
Cascade Range, 72
Cascadia subduction zone, 87
catastrophism, 24, 56, 120, 122
Cenozoic Era, 27, 68, 81, 116, 118
Central America, 87, 132
Chamberlin, T.C., 93, 138, 139, 143
Channeled Scablands, 144
chemical weathering, 81–83, 172
Chicago, University of, 44, 122
Chicxulub crater, 56–57, 61, 120–1
Chile, 84
China, 88, 90, 132
climate change, anthropogenic, 94–95, 128–31, 144–5, 148–58
climate engineering, 15–16, 104, 155–7
climate, factors influencing, 80–83, 99–100, 111–3, 121–2, 134–55
coal, 89–90, 120, 180
Cold War, 14, 52, 120
Colorado, 127
comets , 98, 122
continental crust, 60–61, 63, 71–73
convection, mantle, 64–65
Copenhagen, University of, 130
Corals, 52, 81–83, 167
core of the Earth, 42
cosmic rays, 50–51
creationism, 7–11, 23–24, 174–5
Cretaceous (K-T) extinction, 54–57, 119–22, 124
Cretaceous Period, 39, 73, 81, 121, 159
Croll, James, 136–7, 139
crust of the Earth, 42–43, 60
Crutzen, Paul, 120, 129
Cryogenian Period, 111–114
Curie, Marie, 33
cyanobacteria, 103, 104
Cyprus, 154
Dangaard-Oeschger cycles, 140
Darwin, Charles, 25–56, 28–32, 35, 40, 61, 63, 84, 116–7, 137, 175–6
Dawkins, Richard, 175
day, length of in geologic past, 99
dead zones in ocean, 123–4, 129, 153
decay constant, 33–35
Deccan Traps, 121
decompression melting, 64
Descartes, René, 8
Devonian Period, 35, 40, 91, 123
diamond, 59
dinosaurs, 54, 57, 68, 81, 119
Dobzhansky, Theodosius, 175
"Doomsday vault," 180
Durkheim, Émile, 176
Dylan, Bob, 62, 77, 104

earthquakes, 84–88; human-induced, 99, 150
East African Rift, 53
Ediacaran organisms, 114
Edinburgh University, 43
El Niño, 156
England, 26–27, 29, 40, 90, 114, 144
erosion, 24–25, 62, 75–78, 79–81, 113, 129, 172
erratic boulders, 134–5
Ewing, Maurice, 67

"faint young Sun" paradox, 99
feldspar, 127
Fermi, Enrico, 42–44, 45
fission track dating, 76
fossil fuels, 52, 74, 89–91, 120, 128, 144, 147–154
"fracking" (hydrofracturing), 91, 149, 150
Freud, Sigmund, 176
Future Library Project, 168

Galapagos, 80
Galena, 42
Galileo, 98
Ganges River,
gas hydrates (methane clathrates), 147
Gayanashagawa (Great Law of Peace), 165

geochronology, 36, 40–61
geologic time scale, 8–9, 26–28, 35, 39–40, 68, 130–1, Appendix I
geomagnetic time scale, 68
George, Russell, 153
Glacial Lake Agassiz, 135
Glacial Lake Oshkosh, 159
glaciers, 81, 83, 94–96, 111–2, 124, 130–1, 135, 141, 143–4
Global Boundary Stratigraphic Section and Point, 39, 130
gneiss, 35, 98, 107, 163
Gondwanaland, 73
Grand Canyon, 37, 113
granite, 49, 59, 127
Great Lakes Indian Fish and Wildlife Commission (GLIFWC), 171
Great Oxidation Event, 103–8, 113–4, 152
Great Plains, 81, 110
Great Unconformity, 113
greenhouse gases, 81, 99–100, 111, 121–2, 123–24, 138, 141–55, 172
Greenland, 43, 98, 100–1, 130, 138
greenstone, 99
groundwater, 72, 163–64
Gulf Stream, 143

Hadean Eon, 97
Haida First Nation, 153
Haiti, 85
Haldane, J.B.S., 8
half-life , 34–35, 47–49. *See also* radioactivity.
Hawaii, 80
Heart Mountain landslide, 89
Heezen, Bruce , 67
Hillis, Daniel, 169
Himalaya, 73–77, 81, 83, 112, 153, 172
Holmes, Arthur, 33, 35–36, 40–46, 61, 64
Holocene, 130, 134, 143–4, 158
hot spot, mantle, 80
Hubble, Edwin , 44, 47
Hutton, James, 24–26, 28, 61, 63, 77, 91, 97, 101, 131

Ice Age. *See* Pleistocene.
ice cores, 55, 96–97, 130, 138–41, 168
Iceland, 62–63, 66, 68–69
impact cratering, 56–57, 98
index fossils, 26–28, 116
India, 73, 74, 80, 121
Indian Ocean, 77
Indiana, 38
Indonesia, 72, 84, 85
intergenerational projects, 167–9, 174, 176–177
International Commission on Stratigraphy, 38–39, 57, 61, 130
International Maritime Commission (UN agency), 153
Iowa, 38, 44
iridium, 55–56
iron fertilization of oceans, 104, 142–3, 153
iron formation , 101–2, 108, 180
Iroquois Confederacy, 165, 167
island arcs, 72, 73
isostatic rebound, 83
isotopic dating, 33–35, 40–61
isotopes, stable, 139; of carbon, 100, 139, 146; of hydrogen, 139; of oxygen 60, 139
Isua supracrustal complex, 99–101
Italy, 55

Jack Hills, Western Australia, 58–61, 97, 98, 101
James, P.D., 166
Japan, 72, 84, 85, 87
Joly, John, 30, 71

K-Ar (potassium-argon) dating, 52–53
K-T extinction. *See* Cretaceous extinction.
*kairos*, 26, 61, 168
Katrina (hurricane), 158
Kawara, On, 168
Kelvin, Lord (William Thomson), 28–30, 33, 36, 55, 64, 74, 117

Lackner, Klaus, 150
land plants, in Anthropocene, 130; earliest, 79; in PETM event, 146; role in carbon cycle, 141, 151–2

landslides, 88–89
Laxness, Haldor, 1
Latour, Bruno, 161
lead isotopes, 35, 41–46
lead pollution, 45–47
Leopold, Aldo, 78, 179
life, earliest evidence for, 100–3
limestone, 55–56, 82–83, 101, 122, 144, 153–4
Longyear, John Munro, 180
Longyearbyen, Svalbard, 180
Louisiana delta, 90
Lyell, Charles, 21, 25–26, 31, 61, 63, 79, 89, 94, 95, 97, 131, 134, 135, 137

Madden-Julian Oscillation, 156, Appendix III
magma generation, 64–65, 71–73
magnetic field, 51, 67–68
Manhattan Project, 42, 52, 55
Manitoba, 135
mantle, 42, 63–65, 70–73, 80, 83, 154, 172
Mars, 66, 96, 99, 101, 170
mass extinctions, 54–57, 117–25, 129, Appendix III
mass spectrometer, 41, 59
Matthews, Drummond, 67, 71
Mayan civilization, 132
McPhee, John, 22
megathrust earthquakes, 84
memes, 175, 176
Mesozoic Era, 28, 116
Metamorphism, 10, 53, 57
meteorites, 44–46, 55–57, 119–20, 122
methane clathrates, 147
methane leaks from wells and pipelines, 149
Michigan, 38, 144, 171
microbial life, 103–6, 108–11, 123–6, 141–2, 146, 167
Mid-Continent Rift, 110–1
mid-ocean ridges, 62–66, 103
Milankovitch, Milutin, 136–7
Milankovitch cycles, 136–7, 139–41, 142, 169, 172
Minnehaha Falls, 38
Minnesota, 38, 41, 81, 98, 104, 127, 135, 171
Mississippi River, 38, 91, 144
mitochondria, 105
mixing time (of a geochemical entity), 187
monsoon, 80, 156
Montreal Protocol, 156
Moon, 98–99, 136
Mount Pinatubo, 15–16, 121, 155
Mount Rainier, 60
Mount St Helens, 120, 121, 127
mountain building, 73–75
mountain top removal, 89
Muhammad, 51
Muir, John, 78, 127

NASA, 14
Nazca culture, 131
Neptune, 98
New England, 120, 151
Newfoundland , 114, 154
New Hampshire, 88
New Zealand, 72, 87
Nier, Alfred, 41–43, 61
Nile delta, 90
nitrogen, 50–51, 105, 129
Nobel Prize, 13, 46, 54, 129
Norns (Norse mythology), 162
North Dakota, 135
Northwest Passage, 96
Norway, 2–5, 10, 33, 93–96, 168, 180
nuclear tests, 52, 130
nuclear winter, 120

ocean acidification, 120–1, 123, 149, 155
ocean crust, 62–73
ocean currents, 140, 143, 173
ocean trenches, 69
Ohio, 38, 144
Ojibwe people, 171
Oklahoma, 91
olivine, 154
Oman, 154
Ordovician period, 38, 40, 123
ore minerals, 42, 87
orbital cycles, 136–7, 139–40, 169

*Origin of Species*, 28–32, 116, 117
Ovid, 6
oxygen, rise of atmospheric, 103–8, 113–4, 152
ozone, stratospheric, 105, 156

Pacific Ocean, 68, 69, 132
Paleocene-Eocene Thermal Maximum (PETM), 146–8, 173
Paleozoic Era, 28, 35, 40, 116, 123
paleontology, 26–28, 56, 114–8, 121, 122
palimpsest, 22
Pangaea, 32
partial melting (of mantle), 43, 65, 72
Paterson, Katie, 168
Patterson, Clair, 44–47, 58, 61
peat, 142, 144
pegmatite, 127–8
peridotite, 154
Permian extinction, 123
Phanerozoic Eon, 58
Philippines, 72
Phillips, John, 116–7, 123
phosphorous, 109, 114, 129
photosynthesis, 100–1, 105–6, 108, 120, 130, 144, 145, 150–2
Planktos Corporation, 153
plate tectonics , 63–76, 107, 172
Plato, 13, 167
Pleistocene Epoch, 83, 130, 134–143, 145, 159
potassium-argon (K-Ar) dating, 52–53
Precambrian time, 28, 35, 57–61, 78, 97–115
Proterozoic Eon, 103–15, 127, 152, 163
pseudotachylyte, 86
Public Trust Doctrine, 171
pyrite, 105

Quartz, 127
Quelccaya ice cap, Peru, 131

radioactivity, 32–36, 47–54. *See also* isotopic dating
rare earth elements, 127
reforestation, 151–2
residence time (of a geochemical entity), 187
rivers, 77, 79, 81, 90–91, 105, 112, 144, 172
Rodinia (supercontinent), 112
Ruskin, John, 74–75
Russia, 3, 114
Rutherford, Ernest, 33, 35, 41

Sagan, Carl, 99, 120
St Peter sandstone, 38
*sankofa* (Ghanaian concept) 162, 178
*sati* (Buddhist concept), 162, 178
sandstone, 38, 58, 113
Scheffler, Samuel, 166
Schimper, Karl , 135
Scotland, 24–25, 61, 63, 78, 91, 136–7
seafloor spreading, 65, 67–69, 81
sea level rise, 62, 91, 129
sea sediment cores, 138–9, 146
sea turtles, 69
seawater, chemistry of, 30, 71, 103–4, 108–9, 120, 123–4, 129, 152–3
Secretary of the Future, 173
sedimentary rocks, 24, 36–39, 76, 117, 138–9
seed bank, 180
Sendak, Maurice, 177
Sepkoski, Jack, 122
Seventh Generation concept, 165
Shakespeare, William, 93
Siccar Point, 24–25, 61, 63, 78
Silicon Valley, 170
Silurian Period , 25, 79, 91
slab pull, 69
slow earthquakes, 86
Smith, William, 26, 61, 116
Smolin, Lee, 175
Snowball Earth, 111–4, 128
"solar radiation management," 15, 155–7
Solar System, early history of, 44–45, 98
South America, 69, 84, 131
South China Sea, 90
Sri Lanka, 35
Stonehenge, 167
stromatolites, 101–2, 167
stratosphere, 15, 105, 155

sturgeon, lake, 159–61
subduction, 69073, 84–85, 87
submarine fans, 77, 89
Suess, Eduard, 74
Suess, Hans, 52, 74
Suess effect, 52
sulfate, in atmosphere, 15–16, 121, 155–157
Sumatra, 84
Sun, evolution of, 99, 105, 111, 172
Sunda plate, 85
supernova, 50
Superior, Lake , 103, 111
Surtsey, Iceland, 62–63, 90
Sussman, Rachel, 167
Svalbard, 2–5, 93–96, 124, 128, 141–2, 180
Sweden, 138

Tang dynasty, 132
tektites, 56
terraforming Mars, 170
Ten Thousand Year Clock, 169
Tharp, Marie, 67, 69, 77
thermochronology, 76
Thirty Years War, 133
Tibetan Plateau, 75, 77, 81
tides, 99
tourmaline, 127
Tragedy of the Commons, 166–7
tree rings, 51–52, 59
trench, ocean, 69
Triassic Period,, 123
tsunami, 84, 85
Tyler Prize for Environmental Achievement, 45

U-Pb dating, 35–37, 58–60
unconformity, 24–25, 61, 63, 78, 113
uniformitarianism, 24–25, 61, 78–79, 95, 131, 134–5, 177
universal Darwinism, 175
Universe, age of, 44, 48
uranium, 33–35, 58, 76, 105
Ussher, James, 23

Venus, 66, 96, 99
Vespucci, Amerigo, 66
Vikings, 68–69, 162, 173
Vine, Fredrick, 67, 71
violence, and climate change, 132–3
Vishnu Schist, 113
Vizenor, Gerald, 171
volcanoes, 15–16, 37–38, 62–65, 72, 80–81, 104,113, 121, 144, 172
Vonnegut, Kurt, 173

Wales, 40
Warrawoona Group, 101
Washington State, 87, 144
water cycle, 71–73, 99, 113, 139
West Virginia, 89
*Where the Wild Things Are*, 164
Wieseltier, Leon, 164
Winnebago, Lake, 159–60
winter, 1–2, 159–61
Wisconsin, 1, 38, 78, 93, 110, 128, 134, 137, 143, 151, 159–60, 163, 171
World War I, 32, 133
World War II, 42, 89
Wyoming, 89
wyrd (Norse concept), 162, 164, 178

Y2K crisis, 6
Yellowstone, 50, 88
Ygdrassil, the World Tree, 162
Yosemite, 50
young Earth creationism, 7–11 49
Younger Dryas, 143
Yucatan, 57, 121

Zircon, 9, 33, 58–61, 97, 98